ADDITIVE ALERT!

What Have They Done To Our Food?

A Consumer's Action Guide

Prepared by
Pollution Probe
(Randee Holmes)

Canadian Cataloguing in Publication Data

Holmes, Randee.
Additive alert

4th ed.
Previous eds. written by Linda R. Pim.
ISBN 0-7710-7139-6

1. Food additives – Handbooks, manuals, etc.
I. Pollution Probe Foundation. II. Pim, Linda R., 1952- . Additive alert. III. Title.

TX553.A3P54 1994 664'.06 C93-093191-2

The publishers acknowledge the support of the Canada Council and the Ontario Arts Council for their publishing program.

Typesetting by MACTRIX DTP
Printed and bound in Canada.
The paper used in this book is acid-free.

McClelland & Stewart Inc.
The Canadian Publishers
481 University Avenue
Toronto, Ontario
M5G 2E9

Please note the information in this book is based on research conducted in 1992.

2 3 4 5 00 99 98 97 96

CONTENTS

LIST OF TABLES

PREFACE

Additive Alert was first researched and written by Pollution Probe staff member Linda Pim and published in 1979. Linda updated the book in 1981 and again in 1986. Since then a number of changes have taken place – some additives have dropped out of use (or been banned) while new ones have been added, technological advances have raised new concerns about, for example, microwave ovens and food packaging, and government regulations and industry practices have altered – sometimes in direct response to consumer concern.

This fourth edition of *Additive Alert*, researched and written by Randee Holmes, is the most comprehensive and expanded version of the book to date. Previous editions focused on chemicals deliberately added to food during processing and, while this edition includes the most current information on this subject, in addition, it includes substances that unintentionally appear in our food, such as pesticide residues, packaging chemicals, and veterinary drugs. Finally, the Food Additive Index (abridged on a pocket-sized tear-off card at the back of the book) reflects recent amendments to the federal *Food and Drug Regulations*.

FOREWORD

REFLECTIONS ON THE SAFETY OF FOOD ADDITIVES

by

Dr. Ross Hume Hall

Imagine sitting at the breakfast table and telling your five-year-old, "Go ahead, eat your Crunchies with their 100 deliberately added emulsifiers, acidulants, firmers, enhancers, humectants, sequestrants, stabilizers, texturizers, anti-caking agents, their 150 pesticide residues, their 2,000 petroleum-based residues rubbed off the packaging. Don't worry, because the government food regulations say all those additives are safe." I think your child would be a bit sceptical.

I'm sceptical. Like a five-year-old I keep asking "how can you be sure?" How do you know that the great mix of additives buried in every spoonful of Crunchies is safe?

We have to rely on Health and Welfare Canada to estimate the degree of safety of eating Crunchies and all other additive-containing, heavily-processed foods in your kitchen. And the science these estimates are based upon is, unfortunately, very crude; they can't tell us, precisely, what these chemicals do to us. To begin with, government policy is based on one potentially harmful assumption – what you don't know can't hurt you. If scientific tests don't turn up any evidence of harm, the additive must be safe.

Health and Welfare states its policy in a more neutral, bureaucratic way: "All food additives when used in accordance with provisions of the Food and Drug Regulations are considered to be safe."

The weakness of toxicology tests frustrates scientists. My background is in human biochemistry. For much of my professional career I have studied how foreign chemicals, like food additives and environmental contaminates, affect the body. Such studies are not simple and the results are not easy to understand. I sit on advisory committees for Environment Canada and for Health and Welfare Canada, where we wrestle with the problem of what to do about foreign chemicals people are exposed to when we don't have detailed information on what these chemicals actually do to people. How do we make judgements? As I have said, the government position is: don't judge without proof. Thus, without proof of harm, government policy allows chemicals galore into food. I feel this policy is mistaken.

Why do I feel this way? A short walk through the science of toxicology shows why this discipline does such a poor job uncovering any adverse evidence of health effects caused by the chemicals added to food.

Toxicology is the science that tells you whether a chemical is poisonous – toxic. There are two considerations in judging whether a substance is poisonous: What is the exposure? and, What is the inherent toxicity of the compound? You must be able to answer both these questions before you can judge whether a particular chemical, used in a particular way, poses a threat to human health. Let's start with exposure.

Strychnine is highly toxic, but as long as it stays in the bottle your exposure is nil. If it winds up in your Crunchies, head to the hospital! The word "provisions" in Health and Welfare's bureaucratic statement, "accordance with provisions," defines people's exposure to food additives according to the amounts of foods an average person would eat. For instance, how much low-calorie food containing artificial sweeteners does the average person consume? Food and Drug officials make an estimate. Their estimate doesn't consider people like a former graduate student of mine who drank twenty-four cans of diet cola every day. The government regulations are designed to protect that mythical person – the average Canadian – leaving my student to his chances.

For the sake of argument, let's accept that Ottawa's average person and the amount of food additives she or he consumes is a reasonable estimate of exposure to a particular food additive. Let's now talk about the inherent toxicity of additives. Here's where the science of toxicology falls flat on its face. People, in general, are exposed to low levels of food additives, consumed year after year. Toxicology does a poor job of assessing the risks of consuming small doses of a substance day after day for years.

Toxicology as a science has its roots in the nineteenth century. Toxicologists of that day designed the science to study the effects of high doses of a substance taken over short periods. However, these short-term experiments uncover only knock-you-over-the-head dangers. Nineteenth-century scientists were primarily interested in assessing the safety of the chemicals used in factories. When workers were exposed to high levels of a chemical, some would get sick. So the early toxicologists asked, "Does a short exposure to a particular chemical cause the ill health?" Today, some 150 years later, toxicologists in their laboratories still look only at short-term, high exposures.

Toxicologists have never developed the laboratory tools for measuring the health risk when exposure is low-level over decades. They still dose laboratory mice and rats with large amounts of a suspected toxic compound and watch to see if some evidence of cancer or other pathological effect surfaces during the animals' life-span. Rats have a useful life of about eighteen months; mice live even shorter lives. But rats and mice are poor models for chronic exposure in a human body, which lasts some eighty-five years. In fairness, toxicologists can use long-lived animals such as Rhesus monkeys or chimpanzees. But budget-breaking upkeep, as well as ethical concerns of experimenting on such animals, curtails their use.

The short-term testing of food additives in rodents at large doses picks up only the big dangers. Rest assured, substances like strychnine and arsenic are rejected. But a plethora of nasty substances that the testing misses can get into your food.

Scientists love to talk about what they know. I worry about what we miss, the gaps in knowledge. Our knowledge of food safety is skeletal. And if we don't know very much about the potential for harm from a single chemical additive, what about mixtures?

Nineteenth-century toxicologists addressed the risks of workplace exposure to a single chemical. The science still operates that way today, judging one chemical at a time. Food and drug agencies do not evaluate the additive mix. They make a decision about additive safety one chemical at a time, each decision made as if no other additive ever got into the food. Hardly the case in real life. So what mayhem might the mix of acidulants, stabilizers, pesticide residues, rubbed-off packaging and so on in our morning Crunchies be causing inside the body?

To answer this question we first have to ask what is health? The World Health Organization (WHO) defines health as "a state of complete physical, mental and social well-being, and not merely the absence of disease or infirmity." In WHO's view, a person is entitled to feel good. WHO believes the potential for wellness is built into our genes at conception. From them on, we ought to avoid doing things that knock down that potential.

Government agencies that set policies about the safety of food additives have a lot of trouble with the WHO's concept of what constitutes good health. They leave out the part about well-being. They consider you unhealthy only if you are in a hospital, or under a doctor's care, with a diagnosable disease or condition. You have to be really sick to meet their definition of ill-health. This narrow definition of ill-health has repercussions for you, the consumer. If government officials are unable to connect some serious illness or death to the consumption of a food additive, the additive can be labelled safe.

Government regulators ignore two important aspects of well-being. The first relates to a broad category of poorly defined illnesses. Certain individuals react adversely to chemical exposure, including chemicals used as food additives. They feel nauseated, not at their peak. Or worse, they may be incapacitated, unable to work. Such malaise goes under the name Multiple Chemical Sensitivities. The diagnosis is neither easy nor straightforward. The medical profession has been unable to give the illness sharp outlines. As a result, it falls through the bureaucratic cracks. It isn't considered important in the government's assessment of the safety of a particular food additive.

The second aspect of well-being that often falls through the cracks is chemical assaults on the human foetus. Human beings are extremely vulnerable during those first nine months. Nature designed the placenta without reckoning on human ingenuity in making millions of synthetic chemicals, many of which wind up in our food. The placenta does a poor job of screening foreign substances. The foreign chemicals a mother eats flow largely unimpeded into the foetus she carries. The foetus, unlike the adult, lacks protective, detoxifying systems. It has no protection at all against whatever harm the chemicals might do.

Chemical exposure may cause changes in physical or mental development. Recall the thalidomide story. This drug, taken by the mother during pregnancy, causes a gross birth defect. The embryonic buds that would normally become arms and legs stop developing. The baby is born without limbs. Toxicologists are well aware of such harm caused during pregnancy. They have a name for it – teratogenesis. The thalidomide babies are grossly harmed. Harm can also take subtle forms. A baby, born whole and sound, later in life can experience diminished potential, such as lowered mental agility or behavioural changes.

It is difficult to connect such harm with exposure to specific chemicals. But we know, from studies in animals and in humans, that there are many synthetic chemicals that have the potential to harm the some 240,000 foetuses developing in Canada at any one moment.

Government officials dismiss the risk, saying there is no evidence of a connection between approved food additives and foetal harm. True, there is no direct evidence yet. But the lack of evidence is due to the inability of toxicologists to isolate and understand any direct connections that may exist.

When it comes right down to it, we are still pretty much in the dark when it comes to the safety of food additives. Government policy says a food additive or a mixture of additives is okay until it receives positive proof otherwise. In view of our inability to see clearly the dangers of eating and drinking chemical additives, I believe this is the wrong approach to public safety.

Where does this leave you the consumer?

Additive Alert gives you information for making your own decisions about food additive safety. This book deals with the vast mix of substances added to the food we eat and drink. It takes complex scientific information and applies it to the real world and real people living complex, diverse lives. And the book recognizes the great gaps in our current knowledge about food safety. It points out any disturbing studies that might indicate a problem, and shows you the questionable additives you might wish to avoid. Because what you don't know may harm you – especially your children.

Dr. Ross Hume Hall
Professor Emeritus, Biochemistry
McMaster University
October 1992

INTRODUCTION

This is a book about food additives. A food additive is any chemical, natural or synthetic, that is deliberately used in the growth, processing, storage or distribution of food. Using this definition, food additives can then be grouped into two broad categories. First are those substances that are intentionally added to food during processing and that serve a specific function. *Food-processing additives* include preservatives, sweeteners, colours and flavours. Second are those additives that are intentionally used in the cycle of food production but that unintentionally remain in or on food when we eat it. *Unintentional additives* included in this category are chemicals from food packaging, pesticide residues and hormones and antibiotics from animal products.

This definition of a food additive is much broader than that used by the federal government. In the Canadian *Food and Drug Regulations*, a food additive is defined as "Any substance the use of which results, or may reasonably be expected to result, in it or its by-products becoming a part of or affecting the characteristic of a food."

According to the regulations, the following substances are *not* considered food additives:

salt	spices
sugar	seasonings
starch	flavouring preparations
vitamins	agricultural chemicals
mineral nutrients	food-packaging materials
amino acids	veterinary drugs

There are a number of concerns about the safety of many of these items. Flavouring preparations, such as MSG, and modified starches are associated with many negative effects and have been included in the discussion on food-processing additives in Part One.

In Canada's *Food and Drug Regulations* a distinction is made between "standardized" and "unstandardized" foods. There are approximately 300 standardized foods. These are foods for which Health and Welfare has enforced a standard indicating the ingredients the food is allowed to contain and limits for the quantity of each ingredient. An unstandardized food is any food for which there is not a standard prescribed in the *Food and Drug Regulations*.

The following is a list of some foods for which standards exist. Many of the terms listed below represent whole classes of food. For example, whiskey, rum, gin, beer, cider, and so forth, are included under the heading of "alcoholic beverages."

alcoholic beverages	canned fruit
baking powder	canned beans and weiners
bread	canned vegetables
butter	cheese
cacao beans	chocolate
canned tomatoes	cocoa
canned vegetarian beans	coffee
canned beans with pork	colours –natural

colours – inorganic
colours – synthetic
cracked wheat
cream
cream cheese
dressings
fats and oils (shortening, lard)
flavouring preparations
flour
frozen fruit
fruit flavoured drinks
fruit jelly
fruit juice
fruit jam
ice cream
marmalade
meat and meat by-products
meat derivatives (gelatin, bone meal)
meat stews
milk (cows')
mince meat
pickles
prepared meat
preserved meat
relishes
rice
sausage
seasonings
sour cream
spices
tomato puree
tomato ketchup
tomato juice
tomato paste
tomato pulp
water (mineral, spring)

Food-packaging materials (such as preservatives) and veterinary drugs (including hormones and antibiotics) are not considered by the government to be food additives because they are not *supposed* to be present in food when we eat it. This isn't always the case however. Packaging additives may migrate into food, and testing sometimes indicates the presence of drug residues in animal products. These unintentional or accidental additives are discussed in Part Two. Also discussed in Part Two are the issues of food irradiation and food labelling.

Part Three offers practical information on how to avoid food-processing additives when eating out. Also in this section is an extensive list of sources for more information, including consumer advocacy groups, industry associations and government agencies.

Finally, Part Four offers a complete list of food additives allowed to be used in Canada, with an indication of which may be of the most concern.

PART ONE:

FOOD-PROCESSING ADDITIVES

Food-processing additives are intentionally added to foods as part of processing. They are used to impart particular qualities to foods and are added after the food has been grown, harvested and shipped to the food processor.

UNLESS YOU NEVER eat processed foods (including store-bought bread, milk and cheese) and never eat out at restaurants, chances are that during your lifetime you will consume quite a few of the 350 to 400 processing additives that are allowed to be used in Canada. (This does not include the more than 1,500 food flavouring agents and flavour enhancers, including MSG, which are considered food ingredients, not food additives, by Health and Welfare Canada).

The majority of food additives used today are considered to be non-hazardous; but some are definitely of questionable safety, and others have not been tested well enough for us to know if they are truly safe. It is these additives that you should be aware of and can then choose to avoid.

Food-processing additives fall into two very broad categories. First, there are additives that prevent food from spoiling. They inhibit the formation of moulds, the activity of bacteria and the development of rancidity. Second, there are additives that make food more appealing, in flavour, colour or texture. Sometimes the natural attributes of a food are sacrificed in favour of qualities that suit the food industry, such as uniformity of composition, suitability to manufacturing procedures and extended "shelf-life" (the period over which a food product remains fresh and available for sale). When a food is processed, elements that are lost in the manufacturing process often end up being put back in, except in a different form. For example, many vegetable oils contain tocopherol (vitamin E) and carotenoids (related to vitamin A). These help prevent the oil from becoming rancid. In processing, these naturally occurring compounds are destroyed and synthetic preservatives, such as butylated hydroxyanisole (BHA), are substituted.

BRINGING UP BABY

As a rule, food additives are not allowed in foods intended for consumption by infants under one year of age. According to *Food Additives: Questions and Answers*, a booklet prepared by the Health Protection Branch of Health and Welfare Canada, the following exceptions apply:

- ascorbic acid (vitamin C), used in dry cereals containing banana to prevent browning;
- soya lecithin, used in rice and wheat cereal to prevent sticking during manufacture;
- citric acid, used in fruit desserts to shorten the heating process;
- certain food additives essential in manufacturing infant formulas;
- bakery products labelled or advertised for consumption by infants.

Most manufacturers have voluntarily altered their formulas to reduce or eliminate salt, sugar and monosodium glutamate (MSG). None of these substances are considered to be food additives under Canadian law. Similarly, modified starches are not classified as food additives and do not fall under the restrictions. There are twenty-three chemicals permitted as starch-modifying agents. Parents wishing to avoid these substances should watch for the presence of modified starch.

It is important to distinguish among different uses for food additives. Some uses are utilitarian: Adding preservatives to a food because refrigeration is not available is one example. Some uses, however, are questionable. For example, some manufacturers add preservatives to certain foods while other manufacturers of the same type of foods do not. This suggests that the use of the preservative is not really necessary. Another unnecessary use of additives is for cosmetic or aesthetic effects. This includes adding artificial colour to cheese and to breakfast cereals.

Some additives are added directly to the food item. Others become part of the food when it is processed. Processing additives perform many functions, outlined below.

To make our discussion a little easier, we will discuss food processing additives using the following six categories:

COLOURS, which make food look more appealing.
PRESERVATIVES, which keep food from going bad.
FLAVOURS AND FLAVOUR ENHANCERS, which make food taste better.
SWEETENERS, which make food taste sweeter.
TEXTURE AGENTS, which give food a pleasing texture.
PROCESSING AGENTS, which make processing easier.

Processing additives are used almost exclusively in processed foods. They are not added to fresh fruits and vegetables (with the exception of those oranges that have colour added to their skins to make them appear more orange), fresh cuts of meat or fish, or fresh eggs or milk. However, you will find a host of additives in frozen vegetables packaged in sauces, candied fruit, luncheon meats, dried eggs and flavoured milks. Although there are exceptions, generally the more processed the food, the more food additives it contains.

FUNCTIONS OF FOOD ADDITIVES

- anticaking agents, free flowing agents
- antioxidants
- colours, colouring agents
- curing, pickling agents
- dough conditioners
- drying agents
- emulsifiers
- enzymes
- firming agents
- flour-treating agents
- formulation aids
- fumigants
- humectants, moisture retention agents, and antidusting agents
- leavening agents
- lubricants, release agents
- non-nutritive sweeteners
- pH control agents
- preservatives
- processing aids
- propellants, aerating agents, gases
- sequestrants
- solvents and vehicles
- stabilizers, thickeners
- surface-active agents
- synergists
- texturizers

SOME MISGUIDED NOTIONS

All additives are dangerous.

The majority of additives are considered to be non-hazardous. This means that neither short- nor long-term health effects have been observed in even the most susceptible individuals when these additives have been consumed either in large quantities over a short period or in lesser quantities over an extended period.

We need food additives.

Food additives are often used to replace natural elements that are destroyed during processing with their chemical equivalent. Fresh fruits and vegetables are typically rich in colour and flavour. During processing this richness fades, and artificial colouring and flavouring agents may be used as substitutes. (Colourings and flavourings are not allowed to be used in this way for canned fruits and vegetables.) Instead of using additives, why not change the processing methods or offer foods that are less processed? As it is, we are replacing what has been lost instead of preventing it from being destroyed in the first place. It is a little like continually mopping up the mess on the floor from an overflowing bathtub instead of turning off the taps.

Similarly, instead of using additives to provide foods with artificially long shelf-lives, why not sell them with their shorter, natural, shelf-lives? As consumers we could learn to make more frequent, smaller purchases of unprocessed, fresh foods before they go bad, avoiding the need for additives.

Food additives will help to solve the world food crisis.

Food additives do not serve to increase the world food supply. They are used almost exclusively in processed foods consumed in industrialized countries.

Dangerous chemicals exist naturally in some plants we eat. Why worry about food additives?

Poisonous mushrooms, toxic rhubarb leaves and cyanide in apple seeds are just a few of the hazards naturally present in plants we eat. With all these natural risks afoot, why deliberately expose ourselves to more? The presence of one risk does not justify the addition of another.

If I eat an unsafe additive, I am instantly in danger.

Just as you are unlikely to get cancer from smoking one cigarette or liver disease from consuming one alcoholic drink, you will likely not be harmed from ingesting a questionable additive on a limited number of occasions (unless you are allergic to the additive). Most additives would have to be consumed in large single doses in order to produce immediate toxic effects. But if you consume that additive over a long period, the possibility of harmful effects may increase.

Harmful effects do not always show up immediately. In fact, there may be no evidence of anything unusual until weeks,

How appropriate is it to base the safety of human consumption of food additives on tests done on laboratory animals fed megadoses of these chemicals?

There may be a wide margin of error between the effects of additives on animals and on human beings. We may be more – or less – sensitive than test animals to these chemicals, or we may react in very different ways entirely. But at present, there are really no practical alternative toxicity tests that would allow us to eliminate high-dose animal studies.

It is generally considered unethical to experiment directly on human beings (although some people argue that this is precisely what is happening with regard to the suspicious additives already on the market). Others consider it unethical to experiment on animals to protect human interests. Experiments are conducted on animal species that are as closely related to humans as possible, in hopes that health effects will be similar. Hence, tests on mammals such as rats, dogs or monkeys are more predictive (but also more expensive) than tests on bacteria, such as the Ames carcinogenicity test.

To conduct a scientifically valid study by feeding animals the same dose of an additive as people might consume over long periods would require too many animals, too much time and too much money. Instead, large doses of an additive are administered to the animals. With a built-in safety margin, extrapolations are made as to the expected effects of lifetime exposure to a much lower dose.

months or even years after a chemical has been encountered. By then it is virtually impossible to link a miscarriage, the birth of a deformed child, or the development of a disease later in life with your consumption of a food additive as a teenager, for example.

The International Agency for Research on Cancer has estimated that 80 to 90 per cent of all cancers are caused by environmental, rather than genetic, factors. Although for humans the link between eating food chemicals and cancer remains unclear, a number of food additives have been linked with cancer in laboratory animals. If a chemical is proven to cause cancer in one species of animal, scientists assume that it is, at least theoretically, capable of causing cancer in all life forms, including humans. Of course, the risk to any one person is likely small. But even a small risk can become significant when it is multiplied over the millions of people exposed to the substance.

Food additives are only some of the thousands of chemicals we encounter in our daily lives. It is possible that interactions between these chemicals pose risks of which we are completely unaware. When two or more substances combine to produce heightened effects this is referred to as synergy. The synergistic effects of food additives with each other, with food ingredients and with other chemicals we encounter in the environment are almost entirely unknown.

Nothing should be allowed in our food supply that has not been proven to be absolutely, unquestionably safe.

In truth, nothing can be proven to be safe with absolute certainty. There are no guarantees that anything will be trouble-free for all people all of the time. But we can and should expect careful and thorough research to be conducted to ensure that, to the best of our current knowledge, a chemical does not pose a hazard to our health.

ADDITIVE-FREE

You may be surprised to find that some foods are freer of additives than might be expected. Tomato ketchup seldom contains any added colours, flavours or preservatives. Neither do most jams and jellies. (However, ketchup may contain as much as 30 per cent sugar, and jams or jellies as much as 60 per cent, which many would consider a drawback.)

FEDERAL REGULATION OF FOOD ADDITIVES

In 1964 it became a legal requirement that all proposed additions to the food additive tables be subjected to rigorous tests. Food additives in use before this legislation was put into effect have not necessarily undergone such testing. The Health Protection Branch of Health and Welfare Canada is responsible for regulating food additives, and the information that follows is adapted from their booklet, *Food Additives: Questions and Answers*.

All additives permitted in food sold in Canada are listed in fifteen tables in the *Food and Drug Regulations*. When manufacturers want to use an additive that is not yet approved for use in Canada, or to use an approved additive for a different application than the one legislated, they must submit to Health and Welfare Canada:

- data on physical and chemical properties of the product
- reasons that justify its use
- data on amounts to be used
- detailed results of tests made to establish the safety of the food additive under the recommended conditions of use (tests must be carried out on at least two animal species and must include biochemical and physiological tests, subacute and chronic toxicity studies, and reproductive studies)
- data to indicate residues that may remain in the finished food

- proposed maximum limit for residues of the food additive in or upon finished food
- samples of proposed labelling
- samples of the food additive in the form in which it is proposed to be used, of the active ingredients and, on request, of the food containing the additive

This information is evaluated by Health Protection Branch scientists and, if found acceptable, the Food Additive Tables may be amended for this particular request. The additive must not only meet the safety criteria, but must also serve a useful purpose, without disguising a faulty manufacturing process or deceiving consumers about the food's quality. The large costs involved in this process mean that, generally, only one or two additives are added to the tables in any one year. The approval of existing food additives for new uses is more frequent.

Some people have expressed concern that the required tests are carried out by the food manufacturer – the very body proposing to use the new chemical. Others are satisfied with this arrangement, as long as the tests follow strict standards and are reviewed by competent, unbiased experts.

GOOD MANUFACTURING PRACTICE

In some cases the amount of certain food additives that the government allows in foods is stated in "parts per million" (ppm). In many cases, however, and particularly in the case of food colourings, the most guidance they give in determining an allowable amount is with the term "Good Manufacturing Practice." This means that the amount to be used is up to the manufacturer. According to the *Food and Drug Regulations*, this amount "shall not exceed the amount required to accomplish the purpose for which that additive is permitted to be added to that food." In other words, manufacturers may not use more than they decide they need to use.

When a decision is made to permit the use of a chemical in food, a safe level for that chemical is then established. Since humans are often more sensitive to foreign chemicals than are test animals, a margin of safety of 100 is used, meaning that no more than 1 per cent of a dose causing no effect in the test animals may be used in food.

Canadian food regulations state specifically both which foods may contain a certain additive and the level permitted. This level is typically expressed in parts per million (ppm) parts of food or as Good Manufacturing Practice.

The Health Protection Branch has inspectors stationed all over the country who do regular, unannounced checks of all food-processing plants to see that they are using food additives within the permissible limits, and who patrol border crossings to check the additives in food being imported.

The World Health Organization (WHO) regularly issues bulletins relating to the safety of selected food additives. It is up to each country to weigh its own data on a specific food additive against the WHO's data, and decide whether to continue to permit its use.

1

COLOURS

COLOURS ARE ADDED to foods for no other reason than to make them look good, as the way a food looks affects our perception of how it tastes. In a study done in the early 1970s, volunteer subjects were served a meal under lights that hid the fact that the colours of the foods they were eating had been changed. When, under normal lighting, the subjects discovered that they had been eating blue steak, red peas and green french fries, some of them actually became ill.

Manufacturers believe that food colourings are necessary because we, as consumers, expect foods to be certain colours. And this is generally true. But in some instances this expectation has been learned. We would never expect maraschino cherries to be brilliant red or green or Cheddar cheese to be bright orange if the colours that give them these unnatural hues had not been added in the first place. If we have learned to expect strawberry jam to be a rich and vibrant red, could we not also learn to love a version that, taste unchanged, was not quite so bright?

As with most food additives, colours are used almost exclusively in processed foods. (Colour is not permitted to be added to unprocessed meat for any reason. Processed meats, however, often have nitrites added, which serve to preserve the meat and

Why do butter wrappers state that the butter 'may contain colour'? Don't the dairies know whether they put in colour?

The natural colour of butter varies with the time of year. In the summer, the cattle are grazing in open pasture. Their butter is the characteristic yellow colour because of the presence of carotene pigments in their diet. By contrast, winter butter is much paler because the feed eaten in that season is not as high in carotene. So it is primarily in the winter that extra colour is added to butter. To avoid having to change wrappers with the change in seasons, the dairies cover both possibilities by stating that the butter "may contain colour."

Although the most commonly used colour, annatto, is from a natural source and is purportedly non-toxic, some thirty other colours are allowed, including some synthetic ones of questionable safety (such as caramel, amaranth and tartrazine). Without contacting the dairy directly, we can't be sure which colour is being used, since the name of the colour does not have to be listed on the wrapper.

also bring out a pinkish colour.) Food processors often add colours to their products to replace some of the natural colour lost during processing and storage. Colours are used in artificial foods to make them appear more like their natural counterparts. For example, colour is added to powdered drink crystals to make it resemble real juice. Sometimes colour is added to make the food appear to be more nutritious than it actually is. For example, yellow colour may be added to some bakery products to make them appear to be rich in eggs. Brown colour is added to some bread to make it darker and to suggest that it is a whole grain product.

Food colours are of two types: natural and synthetic. Natural pigments and extracts come from plants, animals and minerals.

COCHINEAL

Cochineal is a naturally derived colouring that gives a purply-red hue to foods. It may be added to processed meat products, condiments such as pickles and relishes, margarine and butter, juices and jams and jellies. Cochineal is made from the dried bodies of the female insect *Dactylopius coccus*, among other species, which are pulverized and boiled in water. The resulting extract, cochineal, is then concentrated. According to *Hippocrates* magazine (July/August 1989) cochineal has been highly valued for hundreds of years: "Indeed, when 16th century Spanish conquistadors demanded tribute from the Indians of the Americas, a pound of the pigment could be substituted for a pound of gold." Between 50,000 and 75,000 insects are needed to make 0.4 kg (1 lb) of the food colouring.

They include annatto (from seeds), turmeric (from a herb), red beet juice, carmine (from dried insect bodies), grape-skin extract and caramel. Newer sources include red cabbage, carrots, cranberries and gardenias. But the majority of colours used in food today are synthetic dyes. Synthetic dyes are made from aromatic amines, which are derived from coal tar and petroleum. Food dyes made from coal tar have been linked to allergies, cancers, hyperactivity and behavioural and learning problems in children.

Why do dyes whose safety is suspect continue to be used? It can be explained as the fast-food philosophy.

Much of the success of fast-food restaurants stems from the fact that customers know exactly what to expect. The food looks, tastes and usually costs the same every visit and in every location. It is predictable. It is for many of these same reasons that food manufacturers prefer synthetic over natural dyes. A dye manufactured in a laboratory can be reproduced exactly. The food manufacturer knows precisely how much dye to add to get the desired colour and to meet the consumer's expectations. And the cost of the dye stays relatively stable over time. Natural colours are usually more expensive to use than synthetic dyes, and they are subject to the same price fluctuations as any food crop. Poor weather conditions and diseases can drive prices up even further and restrict availability. As well, natural dyes are usually not as intense in colour as the artificial dyes and often fade when exposed to light. They may also contribute undesirable flavours to the food. This is particularly true for colours derived from beets and cabbage. Natural colours also tend to be less stable than synthetic dyes in acidic foods.

It is dangerous to assume, however, that just because an additive is derived from a natural source it is therefore safe for human consumption. In fact, many natural dyes have not undergone sufficient toxicity testing. Since their sole purpose is cosmetic, many people feel that colours should not be used at all. In

CITRUS RED No. 2

This synthetic dye was discovered to be toxic in 1973 and was withdrawn from use in all edible portions of foods. However, it is still permitted to be used to colour the skins of oranges. Oranges grown in California naturally have orange skins when ripe. Most varieties of Florida oranges, however, are green when ripe. These oranges are placed in heated chambers, sprayed with ethylene gas and then either sprayed with orange dye while still hot or immersed in vats of hot dye.

The law requires that dyed oranges be labelled as such, but the labels need only appear on the boxes in which they are shipped and not on the oranges themselves. The logic behind allowing the dye to continue to be used in this way, despite the fact that it is a probable carcinogen, is that it does not pose a health hazard unless ingested. So much for orange peel in marmalades, decorative wedges in drinks and using zest when cooking or baking.

**WHAT YOU CAN DO ABOUT . . .
CITRUS RED No. 2**

- Ask your produce manager to display a sign stating whether the oranges are dyed. If the signs give no indication, ask to see the carton in which the oranges were shipped.
- Buy California oranges instead of those from Florida.
- Never use orange peel in recipes unless you are certain that the skins are not dyed. No added colour is permitted on any other citrus fruits besides oranges.

Norway, the use of all synthetic colours in food was banned in 1979. Table 1 describes some of the food colouring agents that are allowed to be used in Canada. All of those listed are considered to be of questionable safety by certain researchers, scientists, physicians and consumer advocacy groups. But don't expect to find any of the food colours listed specifically by name on food labels. By law, food manufacturers need only state "colour" on the label and nothing more than that. It may be wise to avoid foods with added colour.

TABLE 1 Colours of Questionable Safety

The following colouring agents are approved for use in Canada as outlined in the Canadian *Food and Drug Regulations*.

NAME	TYPE	EFFECTS	STATUS	PERMITTED IN
allura red (U.S. red dye no. 40)	synthetic	• may cause cancer in animals	banned in Britain; not approved by the EC*	jam, jelly, bread, butter, fruit juice, ice cream, ice milk, icing sugar, liqueurs, alcoholic cordials, some marmalades, flavoured milk, pickles, relishes, sherbet, smoked fish, lobster paste, fish roe (caviar), tomato ketchup, prepared fish and meat
amaranth (U.S. red dye no. 2)	synthetic	• prevented pregnancies and caused stillbirths in rats • increased a variety of cancers in female rats	banned in Austria, Finland, France (except in caviar), Greece, Italy (except in caviar), Japan, Norway, United States, the former Soviet Union, Yugoslavia	same foods as for allura red

NAME	TYPE	EFFECTS	STATUS	PERMITTED IN
brilliant blue FCF (U.S. blue dye no. 1)	synthetic	• produces malignant tumours at the site of injection and ingestion in rats • may cause allergic reactions	banned in all EC countries, and and in Austria, Finland, Norway, Sweden, Switzerland; FAO/WHO** recommend against its use	same foods as for allura red
caramel	natural	• associated with blood toxicity in rats • inhibited metabolism of vitamin B_6 in rabbits		ale, jam, jelly, beer, brandy, bread, butter, cider, cider vinegar, fruit juice, ice cream, ice milk, icing sugar, liqueurs, alcoholic cordials, malt liquor, malt vinegar, flavoured milk, mincemeat, pickles, relishes, porter, rum, sherbet, smoked fish, lobster paste, fish roe (caviar), tomato ketchup, whisky, wine, wine vinegar, honey wine, prepared fish and meat, liquid, dried, or frozen whole egg or egg yolk, vegetable fats and oils, margarine, cheese

NAME	TYPE	EFFECTS	STATUS	PERMITTED IN
carbon black	natural	• contains a cancer-causing by-product	banned in the U.S.A.	same foods as for allura red
citrus red no. 2	synthetic	• linked to damage of internal organs • induced cancer in animals	banned in Australia and Britain; FAO/WHO recommend against its use; U.S. Food and Drug Administration (FDA) has recommended a ban	skins of whole oranges
cochineal	natural	• caused an abnormal effect on spleen tissue when injected into veins or abdominal cavities of laboratory animals • involved in an outbreak of salmonellosis which killed an infant in a Boston hospital, and made 22 other patients seriously ill		same foods as for allura red; also liquid, dried, or frozen whole egg or egg yolk, vegetable fats and oils, margarine, cheese

NAME	TYPE	EFFECTS	STATUS	PERMITTED IN
cochineal *cont'd*		• research unavailable on long-term effects, reproduction and metabolism		
erythrosine (U.S. red dye no. 3)	synthetic	• determined to be a cancer-causing agent • may interfere with transmission of nerve impulses in the brain • shown to affect thyroid glands of laboratory animals, but not humans • demonstrated adverse effects on blood • may cause gene mutation	banned temporarily in Japan; banned in Norway and the U.S.A.	same foods as for allura red
fast green FCF (U.S. green dye no. 3)	synthetic	• can cause allergic reactions • produces malignant tumours at the site of injection under the skin of rats	banned in Australia and Britain	same foods as for allura red

NAME	TYPE	EFFECTS	STATUS	PERMITTED IN
indigotine	synthetic	• can cause allergic reactions • produces malignant tumours at the site of injection when introduced under the skin of rats	banned in Norway	same foods as for allura red
iron oxide (rust)	natural	• unclear effects	banned in all EC countries	same foods as for allura red; also liquid, dried or frozen whole egg or egg yolk
paprika	natural	• very high doses linked to liver tumours in rats • when injected in cats and mice, capsaicin (ingredients of paprika) slowed heartbeat, lowered blood pressure and inhibited breathing		same foods as for allura red; also liquid, dried or frozen whole egg or egg yolk, cheese

NAME	TYPE	EFFECTS	STATUS	PERMITTED IN
paprika *cont'd*		• when administered by tube, reduced body temperature and raised acid secretion due to irritation		
ponceau SX (U.S. red dye no. 4)	synthetic	• damages adrenal glands and bladders in dogs • causes urinary polyps and atrophy of the adrenal glands in animals	banned in Norway and the U.S.A.; voluntary ban in Japan	fruit peel, glace fruits, maraschino cherries
sunset yellow FCF (U.S. yellow dye no. 6)	synthetic	• causes allergic reactions • causes kidney and adrenal gland tumours in animals • may cause chromosomal damage	banned in Finland, Norway and Sweden	same foods as for allura red

NAME	TYPE	EFFECTS	STATUS	PERMITTED IN
tatrazine (U.S. yellow dye no. 5)	synthetic	• causes allergic reactions; may be life-threatening for individuals allergic to aspirin	banned in Austria, Finland and Norway; recommended ban by the EC	same foods as for allura red
titanium dioxide	natural	• unclear	banned in all EC countries	same foods as for allura red; also liquid, dried, or frozen whole egg or egg yolk, vegetable fats and oils, margarine, cheese
turmeric	natural	• curcumin (component of turmeric) affected body liquids and livers of rats • long-term studies needed		same foods as for allura red; also liquid, dried, or frozen whole egg or egg yolk, margarine, cheese

* EC: European Community, comprising Britain, Belgium, Denmark, France, Germany, Greece, Ireland, Italy, Luxembourg, Netherlands, Portugal, Spain.

** FAO/WHO: Food and Agriculture Organization and the World Health Organization – two United Nations agencies that make recommendations to national governments on food additive use.

2

PRESERVATIVES

According to the World Health Organization about 20 per cent of the world's food supply is lost because of spoilage after it has been harvested. Most people, especially in industrialized countries, cannot live directly off the land, and so some method of food preservation is necessary. Adding chemical preservatives to food is only one method of preventing food from going bad. Other methods include drying, freezing, canning, refrigerating, salting, sweetening, curing, spicing, pickling and fermenting.

When no other means of preservation is suitable, non-hazardous chemical preservatives have a definite role to play. But there are instances where the use of preservatives is of more benefit to the food processor than to the consumer. If a processor has been able to get a deal on a large batch of an ingredient the ingredient may sit around for some time before it is all used. It will therefore require some means of preservation. If ingredients were bought and used while fresh, then the use of many preservatives could be eliminated. And if stock were efficiently turned over, preservatives used to extend the shelf-life of foods would not be needed.

Preservatives are often necessary for foods that must be shipped a long distance to market. One way to cut down on preservative use would be to decentralize food processing. If there

What additives are used in drugs?

Many of the colouring agents, preservatives and other additives used in both prescription and non-prescription medications are the same ones used in foods. However, the use of additives in drugs presents an entirely different situation. Consumers typically have a choice in what foods they eat; they likely have much less choice in what medications they take. In other words, it is much harder to avoid additives in drugs.

In 1985, Canadian manufacturers of non-prescription drugs began to list on labels those additives known to cause allergic reactions. These include, for example, tartrazine (U.S. yellow dye no. 5) and sulphite preservatives. Many consumer groups, particularly those focused on allergies and other food sensitivities, would like to see all additives to drugs listed, regardless of whether adverse reactions are known to occur. Such labelling would help determine whether an individual is sensitive to the drug itself or to only a specific brand that contains certain additives.

If you are sensitive to a certain medication, ask your doctor or pharmacist if they know of an alternative form of the same drug without those additives.

were a greater number of smaller processing plants across the country, rather than a few concentrated in large cities, fewer chemicals would be required to preserve the food as it would not have to travel so far before it reached the consumer. The closer the consumer is to the source of the food, the fewer preservatives are required to keep the food from spoiling. As well, a change in consumer buying habits could reduce or eliminate the need for preservatives. Shopping more frequently and making smaller purchases would mean that foods would not require a long shelf-life.

There are basically two types of preservatives, each added to food for different reasons. Antimicrobials prevent the growth of microorganisms in food. Antioxidants protect the food from going bad or discolouring. Both types give food and food ingredients a longer shelf-life. Their use means that manufacturers can reduce their costs by producing, storing and shipping large batches of food at one time. As well, they can buy large quantities of ingredients when they are cheap and store them until needed. In Canada, approximately fifty preservatives are approved for use. Which preservative or combination of preservatives is used depends upon the food itself.

ANTIMICROBIALS

Antimicrobials are used to prevent the growth of microorganisms such as moulds, yeasts and bacteria. Other ways of dealing with microorganisms include sterilization, pasteurization and cooking (which kill them) and drying, freezing and cold storage (which prevent them from multiplying). But the use of antimicrobials allows for a longer shelf-life than any of these other methods. And they can be used in instances where other processes are ineffective. For example, bottled or canned foods are sterile but only until the container is opened. Once exposed to air, microorganisms begin to grow. This is supposed to be a particular concern with foods packaged in large, family sizes because it is assumed that it will take longer to consume the contents after the package has been opened. A more probable explanation for favouring the use of antimicrobials over other processing methods is that methods such as heating, dehydration and freezing require large and expensive processing facilities, or are not suited to the nature of the food.

Another way to inhibit the growth of microorganisms is to add vinegar (acetic acid) or other acids to foods. Acids must be used in relatively high concentrations in order to be effective. To mask the vinegary taste imparted by the acid, salt or sugar is added. Since it is better we eat foods that are low in salt and sugar, adding acids is not always desirable.

In the interest of technological simplicity, manufacturers most often use chemical preservatives as preserving agents in our foods. Those antimicrobials of greatest concern are the following:

Sulphites

Sulphites are preservatives that stabilize colour, enhance flavour and prevent the formation of bacteria in dried fruits and vegetables, in condiments such as pickles and tomato ketchup, and in beverages. In wines, sulphites are used to prevent the growth of bacteria and yeasts that can turn the wine to vinegar. They help to prevent the deterioration of grapes during transportation and storage and may be used to help preserve the freshness of potatoes peeled and prepared in advance of cooking (for example, french fries in restaurants.)

Most people do not even notice the presence of sulphites in their food. But for others, they are a life or death concern. Some asthma sufferers (estimates range from 5 to 30 per cent) and a small percentage of people without any history of asthma are sensitive to sulphites. Mild reactions to sulphites include weakness, breathing difficulties and hives; moderate reactions include vomiting, diarrhoea, abdominal pain and dizziness; and a severe reaction can result in loss of consciousness and even death. The Food and Drug Administration in the United States reports seventeen deaths to date which could be linked to the consumption of sulphites in either foods or drugs.

Even if you are not sulphite-sensitive, there are other good reasons to avoid these preservatives. Sulphites have caused genetic mutations in microorganisms and tumours in rats. And they are the only food additives currently used that have been directly linked to human deaths.

According to the *Food and Drug Regulations*, the following sulphiting agents are permitted to be used in Canada:

- potassium bisulphite
- potassium metabisulphite
- sodium bisulphite
- sodium metabisulphite
- sodium sulphite
- sodium dithionite
- sulphurous acid

One or more sulphiting agents are allowed to be added to the following foods and beverages:

- dried fruits and vegetables, fruit juices, frozen sliced apples and mushrooms
- beverages (alcoholic and non-alcoholic)
- fresh, peeled potatoes
- grapes
- crustaceans (for example, shellfish)
- glucose solids and syrup and dextrose (used in confectionery)
- jam, jelly and marmalade
- molasses
- gelatin
- mincemeat
- pickles and relishes
- tomato paste, pulp, ketchup and purée

Sulphites may also be added to unstandardized foods, such as snack foods and certain confectionery products, but are not allowed to be added to preparations of meat, fish and poultry or to foods considered as a source of thiamine (vitamin B_1) as sulphites

Are additives permitted in milk?

Vitamins A and D are added, but these are not included under the Canadian government definition of food additives. However, preparations of these fat-soluble vitamins for use in milk are allowed to contain the antioxidant preservatives butylated hydroxyanisole (BHA) and butylated hydroxytoluene (BHT). Additives used in the preparation of some food ingredients, including vitamins, need not be listed. That is, the ingredient is listed on the label, but the constituents of the ingredient are not.

Check with your local dairy to determine if the vitamin preparations it uses contain BHT and/or BHA.

WHAT YOU CAN DO ABOUT . . . SULPHITES

Not many people are allergic to sulphites, but for those who are sensitive, the reactions can be severe. This is especially true among asthma sufferers. If you suspect that you are sensitive to sulphites or are simply concerned about consuming them, here are some things you can do to avoid them:

- Become aware of which foods are allowed to contain sulphites. Look for sulphite-free versions.
- Remember that the ingredients of foods may themselves contain sulphites. If the ingredient is one that is not required to have its components listed on the label, it may indeed contain sulphites. (There is movement within Health and Welfare to require that sulphites be listed on the label no matter their source.)
- Don't eat grapes. Of all foods that may be eaten raw, they are the only one on which sulphites are allowed to be used.
- Avoid alcoholic beverages, imported or domestic, unless they specifically state that they are sulphite-free. (Most don't.) Whiskey, rum and vodka are usually free of sulphites.
- When eating out, choose restaurants which participate in the Allergy Aware program established by the Canadian Restaurant and Foodservices Association. (See Chapter 14.)
- If the restaurant is not a participant of the Allergy Aware program, ask the manager if sulphites have been added at any stage in the preparation of the food. Ask the same question when buying foods in bulk. Avoid the food if there is any doubt.
- Be aware that certain medical drugs also contain sulphites. Ask your doctor how you can ensure that any medication you are taking is sulphite-free.
- Request the pamphlet *Sorting out Sulphites* from Health and Welfare Canada. (Their address is listed in Chapter 15.)
- If you suffer a reaction you believe was caused by sulphites, consult a qualified health practitioner and then report the incident to the Health Protection Branch of Health and Welfare Canada.

destroy this vitamin. They are no longer allowed to be used on fruits (except grapes) and vegetables to be consumed raw (such as in salad bars).

Most packaged foods containing sulphites, with the exception of alcoholic beverages, will list the agents on the label. (The Health Protection Branch of Health and Welfare Canada has recently proposed that alcoholic beverages containing sulphites also be labelled as such.) Be careful though. An ingredient of a food may contain sulphites. While in most cases the components of ingredients must be listed on the label following the ingredient itself, there are exceptions. (For a list of ingredients that do not have to have their components listed on the label, see Chapter 11 on labelling.)

Not all foods allowed to contain sulphites will necessarily contain them, as sulphites are not required to be added. Read the labels before buying.

Nitrites and nitrates

Nitrite is a chemical used to prevent the growth of the bacteria that cause botulism. It is also used to enhance flavour and to preserve the red colour of meats. *Nitrate* is a preservative closely related to nitrite. It is relatively harmless until it is converted to nitrite by bacteria present in foods and in the body. Nitrite, either intentionally added to food or formed in the body, can combine with other compounds to form nitrosamines. And nitrosamines are of great concern. According to Beatrice Trum Hunter, the well-known U.S. food additives writer, nitrosamines have been described as "the most potent cancer-causing agent known to science." She reports in *Consumers' Research* magazine (January 1988) that more than 90 per cent of the more than 300 nitrosamines that have been tested in laboratory animals cause cancer. Hunter also says that, according to data collected from many countries, "fried bacon and beer contribute more volatile nitrosamines to diets than all other food sources combined."

By changing processing techniques, brewers have been able to reduce the formation of nitrosamines in beer. And the use of nitrite as a curing agent in many processed meats, including bacon, has declined in recent years. Many manufacturers now also add ascorbic acid (vitamin C) or erythorbic acid to their products as these acids seem to inhibit nitrosamines from forming. Although the use of nitrites is much reduced, these substances are often still present in foods (a study reported in March 1990 by the Washington-based Center for Science in the Public Interest, found them in 90 per cent of 271 luncheon meats). Even at lower levels, nitrites should be avoided.

Don't be surprised if you hear the claim that nitrite actually improves the safety of the food supply by preventing botulism. It is true that nitrite can prevent the growth of *Clostridium botulinum*, the bacteria that cause this potentially fatal condition. But the availability of nitrite-free brands of processed meats, weiners and bacon suggests that nitrite is not really necessary.

It is important to realize that the amount of nitrite entering our bodies from intentional food additives is much less than the amount we encounter in food, water and air contaminated by fertilizers and car exhaust fumes. Nitrate is present in tap water and in a number of vegetables including spinach, beets, radishes, eggplant, celery, lettuce and the greens of collard, turnip and beets. This is due in part to the widespread use of nitrogen fertilizers. According to Nicholas Freydberg and Willis Gortner, authors of *The Food Additives Book*, "about 80 per cent of the human dietary exposure to nitrites comes from nitrates in drinking water and in foods that have not had any added nitrate or nitrite; about 20 per cent comes from cured food products." However, most of the natural food sources of nitrate also contain vitamin C, which inhibits nitrosamine formation. Nitrite is also produced naturally

**WHAT YOU CAN DO ABOUT . . .
NITRITES AND NITRATES**

- If you eat processed meats look for those that are nitrite-free. Better yet, avoid them altogether. They are usually high in fat and salt.
- If you eat bacon, cook it in a microwave at a low temperature so that fewer nitrosamines form.
- Don't cook with fat drippings from bacon.

in the body at levels ten to thousands of times higher than that found in cured meats and other foods. In any case, given that nitrates and nitrites may be associated with the development of cancer, that long-term studies with experimental animals suggest that they may have some adverse effects on offspring when they are consumed during pregnancy, and that infants are more susceptible to problems from these additives than are adults, there is no justification for intentionally adding nitrite or nitrate preservatives to our food.

ANTIOXIDANTS

When oxygen comes in contact with certain food molecules it changes their shape, structure and behaviour. This is called oxidation. Antioxidants prevent oxidation and are used primarily to prevent oils and fats from going rancid. Without them an unpleasant taste and odour would develop and potentially toxic or cancer-causing substances could form. Antioxidants are naturally present in many fatty foods. However, they are not present in high enough concentrations, especially after processing, to keep the food from going bad over a long shelf-life.

Oxygen also acts on certain enzymes in food and causes processed meats to change colour and sliced fruit and vegetables to turn brown. These changes can be prevented with the use of ascorbic and citric acids.

Like most other additives, antioxidants do not serve a vital health function. Their only effect is to extend the shelf-life of foods from months to years. The safety of the following antioxidants is questionable:

Butylated hydroxyanisole (BHA) and butylated hydroxytoluene (BHT)

BHA and BHT are closely related preservatives used to prevent rancidity in such fatty products as oils, lard, margarine and shortening. They are also used in powdered drink mixes, chewing gum and some meats and in the packages of some breakfast cereals.

More research is needed to be sure about the safety of these antioxidants. Some research has suggested that BHA and BHT may protect against certain forms of cancer. But the International Agency for Research on Cancer, part of the WHO, considers BHA to be a possible carcinogen in humans, and the state of California has listed it as a carcinogen. BHT has been found to cause cancer, birth defects and reproductive failure in laboratory animals and is prohibited as a food additive in England.

Propyl gallate

Used on its own, propyl gallate tends to give foods a bluish or greenish hue. To counteract this, propyl gallate is typically used in combination with one or both of BHA and BHT. Combining them also allows manufacturers to take advantage of their synergistic action, which allows smaller amounts of chemicals to be used.

In 1981 the National Cancer Institute completed probably the most thorough study on propyl gallate to date. Researchers found numerous suggestions of cancer in both mice and rats. Other evidence has suggested that when propyl gallate is used in meats containing sodium or potassium nitrite, it inhibits the formation of nitrosamines (see Nitrites and Nitrates, above).

There are other chemicals or processes available that could make the use of BHA, BHT and propyl gallate unnecessary. Vitamins C and E are safe antioxidants, but they cost more than the synthetically produced preservatives and are not as potent. Freezing foods, replacing oxygen with nitrogen gas and packaging foods in dark rather than clear glass bottles could eliminate the need for antioxidants.

The use of BHA, BHT and propyl gallate introduces small but unnecessary health risks that can and should be avoided. If you find a brand of food that contains any of these substances, look a little further. Chances are that you'll be able to find another brand that doesn't contain them.

Even if all potentially toxic chemicals in food were removed, we may still be breathing polluted air and drinking chemicals in water. What good does it do to solve only part of the environmental health problem?

Reducing our exposure to toxic chemicals is the most important thing. We will likely never achieve zero risk from chemicals in the environment, but that is no reason to stop trying to minimize our exposure to these chemicals by whatever means they enter our bodies. Efforts to rid our diet of harmful substances must be made part of the broader process of cleaning up the environment.

3

FLAVOURS AND FLAVOUR ENHANCERS

STRICTLY SPEAKING, under Health and Welfare Canada's *Food and Drug Regulations*, flavouring preparations are considered to be food ingredients rather than additives. However, since most people think of them as additives, and since much concern and controversy surrounds a few in particular, it seems appropriate to mention them here.

FLAVOURS

In Canada, as many as 1,500 natural and artificial flavours are permitted to be used in our food. That is about three times the number of all the permitted additives combined. Yet they are the food chemicals whose safety we know the least about.

Natural flavours are derived from natural ingredients. They are often processed into concentrated oils and extracts and may be combined with other natural flavours to obtain the desired effect.

Artificial flavours are used in place of natural flavours, which can be expensive and hard to come by. They are synthetically produced and combinations of them are usually required to simulate a natural flavour. A common example is artificial vanilla, or vanillin, which simulates the extract of vanilla beans. Vanillin is produced from a fibre in wood pulp. According to

Hippocrates (July/August 1989), much artificial vanillin has historically come from paper mill effluent (waste water).

If a food contains an artificial flavour, you can usually assume that any natural flavour listed on the label is there only in a token amount. The word "natural" has proven to be a consumer eye-catcher and manufacturers play it up as much as possible. And, according to federal regulations, "artificial flavour" or "natural flavour" are the most specific terms that manufacturers need put on a food label. There are few restrictions as to where they are used and in what quantities.

Over the past fifteen years, some food companies have developed a range of "natural" flavours extracted from natural sources. A product may be labelled, for example, "natural strawberry flavour," implying that the flavour has been extracted from strawberries. Not necessarily so. It may be a mix of chemicals extracted from a variety of plants, none of them strawberries. Since the source of the flavouring is from plants, it can be labelled "natural."

Very few flavouring agents have undergone adequate testing. The government is not as strict in regulating them because they are used in such small quantities and most occur naturally. (Synthetic copies of naturally occurring substances are exactly the same in terms of their chemical make-up and properties.) One estimate states that the average person consumes an ounce or less of artificial flavourings in one year. But here we must be careful. The fact that only a small amount of a chemical is ingested or that it occurs naturally in no way guarantees that it is safe.

Many flavour additives have been linked to allergic or other hypersensitive reactions in humans, including coughing, asthma, hives, heartburn, headaches, migraine headaches and hyperactivity in children. Quinine is a bitter flavouring used in beverages such as tonic water. According to *The Food Additives Book*,

FLAVOUR FROM WHAT?

According to Canadian regulations, only the word "flavour" need appear on a food label to indicate that a flavouring agent has been added to a product. No indication need be given about the source of the flavouring agent. Flavourings may be derived from livestock, plants, yeast, dairy products, eggs and fish. *Tufts University Diet and Nutrition Letter* (August 1990) reports that some flavourings may contain gluten, a substance that cannot be digested by individuals who suffer from a disease known as Celia Sprue. For those who are susceptible, gluten acts like a toxin in their gastrointestinal tracts. Other people might want to avoid flavourings derived from animals or animal products for religious or cultural reasons. Religious Jews and Moslems would want to avoid products made with ingredients that come from pigs. Orthodox Hindus would not want to eat anything derived from a cow. Strict vegetarians would want to avoid anything that came from an animal. As with colouring agents, consumers should be provided with more information on food labels than simply the word "flavour" so that they can make informed choices about the products they buy.

"some evidence that hearing may be impaired from high intakes of quinine, and that the fetus may also be subject to toxic effect, suggests that pregnant women should avoid quinine."

FLAVOUR ENHANCERS

Flavour enhancers are used to increase the strength of a flavour without themselves adding a taste to the food. Although they must be listed on Canadian food labels, there are no restrictions on at what levels and in which foods they are used. And a questionable lot they are. Here are just a few – all of them currently approved for use in Canada.

Brominated vegetable oil

Brominated vegetable oil (BVO) is a combination of bromine and vegetable oil. Flavouring oils are dissolved in BVO and then added to carbonated and noncarbonated fruit-flavoured drinks. The density of the liquid that results from combining the bromine with vegetable oil prevents the flavouring oils from separating and rising to the top of the bottle. It keeps them evenly distributed throughout the liquid. BVO also gives the drinks a slightly cloudy appearance so that they appear more like fruit juices.

Various tests with laboratory animals have linked BVO consumption to fat deposits in the heart, liver and kidneys. It has also been found to cause damage or changes to the heart, liver, thyroid, testicles and kidneys. There is also evidence that BVO accumulates in animal tissue. In 1971 the FAO/WHO Expert Committee on Food Additives recommended that BVO not be used as a food additive. Both Sweden and Britain have banned its use.

Caffeine

Caffeine is a naturally occurring stimulant found in coffee, tea, cocoa, and kola nuts. It can be addictive, and it is intentionally added to cola beverages – about half of all soft drinks.

Most people are aware of the effects of excessive caffeine intake. These include restlessness, irritability and insomnia. But caffeine is also associated with other, more serious, effects. Consumed in large doses, caffeine can stimulate sufficient gastric secretions that peptic ulcers result. It may also aggravate heart disease. (The link between caffeine and heart disease has been a subject of debate for many years.) Of concern to women is the evidence that suggests caffeine may cause fibrocystic breast disease (benign breast lumps), interfere with reproduction and influence foetal development. It has been linked to various birth defects in mice, rats and rabbits and may affect bone growth.

Since caffeine in the body can cross the placenta and reach the growing foetus, pregnant women should avoid consuming it. Women who breast-feed their children will pass on some of the caffeine they themselves consume. Caffeine may be dangerous for children, particularly infants, and should be avoided.

Monosodium Glutamate (MSG)

Monosodium glutamate (MSG) is made up of sodium and glutamic acid. Glutamic acid is an amino acid which, when linked with others, forms protein. Because the glutamic acid in MSG is not bound to other amino acids it is called a "free" glutamate.

Various forms of glutamic acid, called glutamates, occur naturally in a variety of foods including meat, fish, milk, cheese, wheat, nuts and vegetables such as onions, mushrooms, tomatoes and peas. They are also naturally found in the human body. The amount of glutamate that we consume daily as MSG is estimated to be about 1/1000 of that naturally present in the body.

Food processors and cooks alike use MSG to enhance the flavour of food. But for manufacturers it has another advantage. Adding MSG can be a way to cut down on other, more expensive, ingredients. It is possible to decrease the amount of fish, meat or vegetables in a product and increase the amount of MSG to maintain the same flavour. MSG is an inexpensive ingredient and there are no federal restrictions on the amount that may be used.

Many people have reported adverse reactions to MSG including "Chinese Restaurant Syndrome." Symptoms include a feeling of tightness in the chest, flushing of the face, headache, and a tingling sensation in the arms. But according to the Institute of Food Technologists (IFT) the association of MSG with adverse reactions has not been clearly established. Although the IFT admits that some people have reacted negatively to foods containing MSG, they are not convinced that MSG itself is the culprit. The IFT states that scientific studies over a number of years have shown that MSG is safe for the majority of people, even when consumed in large amounts. The American College of Allergy and Immunology agrees. Ronald A. Simon, M.D., of the Scripps Clinic Research Foundation in California, suggests that people who report reactions to MSG are prone to those symptoms anyway. That is, if you suffer from migraine headaches you may well get a headache after consuming foods high in MSG. The United Nations' Joint Expert Committee on Food Additives declared in 1987 that there is no link between MSG and after-eating discomfort.

Other people, including George Schwartz, M.D., author of *In Bad Taste: The MSG Syndrome*, are far from convinced. Dr. Schwartz reports cases of a young boy who couldn't control his bowels and displayed hyperactive-like behaviour, a man whose dizzy spells became so severe that he could no longer drive, and a woman who suffered abdominal pain and depression. Another woman had such a severe asthma attack that she had to be put on a ventilator and underwent a cardiopulmonary bypass. In all of these cases, MSG exposure was determined to trigger the adverse effect.

John Olney, a neurophysiologist at Washington University, found in 1969 that a single dose of MSG raised glutamate levels in the blood and caused damage to the hypothalamus, a region in the brain, when given orally to rats and monkeys. His research revealed that glutamate stimulates nerve cells in the brain. He also discovered that it acts as an excitotoxin; when present in excessive amounts it can actually excite the cells to death. Olney thinks that children and especially infants may be at particular risk from MSG in foods and has been fighting for stricter regulation for more than twenty years.

NO EFFECTS FROM MSG?

R. A. Kenney, Ph.D., a physiologist retired from the George Washington University School of Medicine and Health Sciences, believes that Chinese Restaurant Syndrome doesn't exist.

Dr. Kenney, whose research is often funded by the Glutamate Association in Atlanta, set out to prove his point by studying people, who had previously identified themselves as MSG responders, in a double-blind, placebo-controlled study.

Study participants drank a beverage containing either a high dose of MSG or a placebo. Ironically, MSG symptoms were reported more often by people who drank the orange or tomato juice placebo than a solution of MSG.

Dr. Kenney did note that some individuals were hypersensitive to MSG when it was given in large doses. But he pointed out that using such large doses in cooking would be unlikely. Like salt, the taste of MSG has a self-limiting characteristic: A little improves taste, but a lot ruins the flavour.

Excerpted from: Ternus, Maureen. "MSG: still making headlines." *Environmental Nutrition*. Vol. 14, June 1991, p. 2.

In Health magazine (November/December 1990) cites a recent study by the WHO in which findings from twenty years of research on MSG were reviewed: "They acknowledged that rats and mice injected with MSG show brain damage, and they recounted John Olney's more worrisome discovery of brain injuries in monkeys given comparable doses." But nineteen other laboratories were unable to replicate Olney's results. WHO researchers concluded that the consumption of MSG, even in large amounts, causes no harm.

Although there may be little "scientific evidence" that people react to MSG, for those who identify themselves as sensitive, the symptoms they experience after ingesting even small amounts of the substance (including migraine headaches, nausea, vomiting, diarrhoea, mood swings, aching joints and dizziness) are

WHAT CAN YOU DO ABOUT . . . MSG

According to a survey conducted by the Quebec-based *Protect Yourself* magazine, the foods most likely to contain MSG are those packaged in cans or envelopes and frozen foods. Of the foods they studied, canned and dehydrated soups, Chinese food and seasoned dishes contained MSG in almost all cases. If you think that you are MSG-sensitive, start reading those food labels: the magazine reported finding more than 500 products that contained MSG in a single grocery store.

Other ways to avoid MSG: Buy fresh ingredients. Only processed foods contain added MSG. Watch out for MSG in disguise. The following food ingredients may contain glutamate:

- flavourings
- hydrolyzed plant protein
- hydrolyzed vegetable protein (HVP)
- kombu extract
- natural flavourings
- potassium glutamate
- seasoning
- spices
- vegetable, chicken or beef broth

According to *Environmental Nutrition* (June 1991), hydrolyzed proteins made from meat and vegetable sources may contain 8 to 40 per cent MSG. (Steve Taylor, Ph.D., at the University of Nebraska suggests that people who react to hydrolyzed proteins may be reacting to the protein source, such as soybeans or wheat, rather than to MSG.)

- Canned gravies, stews, sauces, fish and meat frequently contain MSG. Read the labels carefully. Also scrutinize the labels of sausage or luncheon meats purchased at the delicatessen or local butcher shop.
- Many packaged dried foods, such as noodles, rice and soups, come with flavour packets. When checking out the ingredients on the package, be sure to read two lists: one for the food itself and one for the flavouring. Almost all of these flavourings contain MSG.
- When eating out, ask if MSG has been added to the foods at any stage in preparation. Be aware that soups served in restaurants may contain higher levels of MSG than similar prepared versions sold in grocery stores.

proof enough. And there is some evidence to suggest that MSG may play a part in asthmatic reactions. But since the symptoms may be delayed as much as twelve hours, making the connection can be difficult.

Estimates of the number of people sensitive to MSG range from a very conservative 1 per cent of the population to as high as 25 per cent. If you think you are sensitive to MSG, you may want to call an allergy association. (See Chapter 15.) You may also wish to contact the National Organization Mobilized to Stop Glutamates (NOMSG)in the U.S.A. You can call them toll-free at 1-800-288-0718 or write them at P.O. Box 367, Santa Fe, New Mexico, 87501. Founded in 1989, the publish a quarterly newsletter which goes out to their approximately 800 members.

4

SWEETENERS

OUR DESIRE TO HAVE our cake and eat it too, with few, if any, calories, has resulted in a huge market for artificial sweeteners. According to Arthur D. Little, a market research company, sales of diet soft drinks in Canada are currently increasing by more than 10 per cent each year. Artificial sweetener sales are expected to grow at a rate of more than 5 per cent each year at least until 1995. In a world market that currently amounts to about $1 billion, that's a whole lot of (artificial) sweeties. And a whole lot of controversy, too.

Aspartame

About 200 million people worldwide use aspartame (trade name NutraSweet). According to a study of more than 10,000 Canadians reported by Health and Welfare Canada, approximately half of all households use artificial sweeteners either in commercial food products or as tabletop sweeteners. Every year Canadians consume $25 million worth of the sweetener aspartame. And Canadian sales make up only 5 per cent of its worldwide market. The United States, where aspartame sells for double the Canadian price, takes care of 75 per cent of the world market.

Discovered in 1966, aspartame was not approved for use in

Canada until 1981. By the end of 1982, twenty-four countries had approved it as a substitute for sugar. Prior to its approval, aspartame underwent more testing than any food additive before it. And yet the controversy over its safety rages on even today.

Two hundred times sweeter than sugar, aspartame comprises the two amino acids phenylalanine (about 50 per cent) and aspartic acid (40 per cent), and methyl alcohol (10 per cent). Although phenylalanine is an essential nutrient, for people with a rare genetic disorder called phenylketonuria (PKU) it can be deadly. One in 20,000 babies is born with PKU, which leaves them unable to metabolize phenylalanine. If not detected early and if too much phenylalanine accumulates in the blood of a child, it may result in brain damage and mental retardation. Because of this, all foods in Canada containing aspartame must state on the label "contains phenylalanine."

Aside from those people with PKU, a number of others (estimates range from 1 to 10 per cent of the population) may be carriers of the genetic trait. These people have only one copy of the gene needed in order to break down phenylalanine instead of the usual two. Phenylalanine in the blood of these carriers can reach higher than normal levels for a given dose of aspartame and may slow their brain waves. Some scientists feel that if female carriers consume large amounts of aspartame and other sources of phenylalanine while pregnant, their babies might be born mentally retarded. Others disagree.

The following table is a summary of the many different arguments put forward in support of the safety or danger of aspartame. Don't expect the debate to be resolved in the near future. You'll have to weigh the evidence yourself and make your own decisions about whether you are willing to submit to the risks, if you believe there are any, associated with aspartame. Better safe than sorry?

TABLE 2 The Question of Aspartame's Safety

ON THE ONE HAND	ON THE OTHER HAND
The Canadian government agrees with the U.S. Centre for Disease Control and the U.S. FDA Advisory Committee on Hypersensitivity to Food Constituents that although some people may have an unusual sensitivity to aspartame, there is no evidence that aspartame usage poses a significant health hazard to the general population.	In the late 1960s and early 1970s, laboratory animals given large amounts of aspartame developed brain tumours. Aspartate, a chemical found in aspartame, was found by Dr. John Olney at Washington University to have the potential to cause brain damage in animals. Other studies suggest that it may affect brain chemistry.
G. D. Searle, the creator of aspartame and initial manufacturer of NutraSweet, submitted studies to the U.S. FDA proving its safety. Some scientists in the agency had reservations about the quality of the studies and the results. The FDA went ahead with its approval. An independent review agreed with the FDA's decision.	Steven Farber, in *Technology Review* (January 1990), reports that a review of Searle's studies by FDA Bureau of Foods scientists revealed significant errors in the data, results that seemed to have been falsified and questionable laboratory practices.
When aspartame was approved, Health and Welfare set the acceptable daily intake (often referred to as ADI) at 40 mg/kg of body weight. This level is in agreement with that recommended by the Joint FAO/WHO Expert Committee in 1980. A food intake survey done in the	Acceptable consumption levels were set when aspartame was initially approved and were based on estimates of how much people would be consuming. Some have suggested that consumption of 50-100 mg/kg of body weight per day might not be uncommon.

ON THE ONE HAND	ON THE OTHER HAND
United States, Germany and Finland found that in all populations and all age groups actual consumption was well below the acceptable limit.	
In one experiment, subjects were given an amount of aspartame estimated to be thirty times greater than the average daily intake. No negative short-term effects of any sort were observed.	According to Dr. H. J. Roberts, Director of the Palm Beach Institute for Medical Research in Florida, more than ten thousand people have reported side effects after consuming aspartame.
A reviewer criticized studies done by Dr. Roberts because he administered a questionnaire to people who thought they were sensitive to aspartame without also polling people who thought they did not react to the sweetener or people who did not use aspartame at all.	Dr. Roberts studied a group of more than 500 people who reported symptoms associated with aspartame consumption. These symptoms include headaches, seizures, dizziness, decreased vision, confusion, memory loss and severe depression.
Some studies have found that subjects taking a placebo were as likely to report headaches as people taking aspartame.	Headaches are the most common complaint associated with aspartame reported to the FDA.
Data from studies suggest that at current consumption levels aspartame does not cause seizures.	Animal studies in several laboratories have found that high doses of aspartame can increase the frequency of seizures or lower the amount of stimulation necessary to induce them. It is not clear, however, that aspartame is responsible for *causing* the seizures.

ON THE ONE HAND	ON THE OTHER HAND
The two amino acids that make up aspartame (aspartic acid and phenylalanine) naturally occur in many foods. According to advertisements, the components of aspartame are the same to your body as those found in natural foods.	Phenylalanine is a component of all proteins. In foods, the presence of other amino acids inhibit the body from taking it up. In aspartame there is nothing to prevent this from happening. Excessive phenylalanine in the body may be responsible for many of the complaints associated with aspartame consumption.
One of the components of aspartame is methanol (wood alcohol). In excessive amounts methanol is poisonous and can cause blindness. Methanol is naturally present in many foods, such as juice, fruits and vegetables. The U.S. FDA suggests that consumers are exposed to higher levels of methanol from these sources than they are from aspartame.	Critics have suggested that there is less methanol in fruit juice than in diet soft drinks. And when fruit juice breaks down to form methanol it also produces ethanol. Ethanol protects against methanol toxicity. Most foods also contain folic acid, another protector against methanol poisoning. Methanol further breaks down into formaldehyde and formic acid. Formaldehyde is a known carcinogen and has produced nasal tumours in laboratory animals. Some feel these substances are responsible for the damage reportedly associated with aspartame.
Billions of dollars are spent by consumers on foods and beverages sweetened with sugar substitutes. Many of these people believe that consuming products sweetened with aspartame will help them lose weight without giving up sweet food.	Numerous studies suggest that aspartame does not help people control their weight. There is even evidence to suggest that aspartame might stimulate appetites, causing people to consume more calories and actually gain weight.

SUCRALOSE

Watch out aspartame – there's competition afoot. Sucralose, which is made from sugar, is the newest artificial sweetener on the Canadian market. It is being marketed under the brand name Splenda by Redpath Industries, a subsidiary of the British company Tate and Lyle.

Approved in September 1991, sucralose looks, tastes and performs much like sugar. But it is 600 times sweeter than sugar and therefore a much smaller amount is needed to simulate the taste of sugar. Sucralose is calorie-free. It does not break down when heated and is therefore suitable for a wider variety of products than is aspartame, which is heat-sensitive and so cannot be used in cooking or baking. Products containing sucralose display an identifying logo.

Just as we were told aspartame had been, sucralose has been more thoroughly tested than any other food additive in history. And also like aspartame, there still remain questions about its safety.

At a cost of $300 million, more than 100 long- and short-term scientific studies were conducted over fifteen years to demonstrate the safety of sucralose. In 1990, the WHO/FAO Joint Expert Committee on Food Additives declared this new sweetener safe for human consumption. But the Center for Science in the Public Interest (Washington, D.C.), and the Scientific Committee for Food (Europe) think that more testing needs to be done. Studies have shown that rats fed diets high in sucralose had shrunken thymus glands and enlarged livers and kidneys. The thymus gland is an important part of the immune system. There are also concerns about a breakdown product of sucralose which may cause mutations.

5

TEXTURE AGENTS

TEXTURE AGENTS modify a food's texture in order to give it a desired consistency. Some foods, such as flavoured gelatin desserts, would not be possible without them. In other cases, texture agents are added for convenience so that we don't need, for example, to shake a bottle of oil-and-vinegar salad dressing before pouring it. Texture agents may also be used to imitate the natural texture that would result if more time and care were taken in processing. Thick yogurt, for example, can be produced without the addition of gelatin, but it is a delicate procedure which requires that the bacterial culture not be disturbed.

Texture agents include:

Emulsifiers to prevent the separation of liquids, such as oil and vinegar in a salad dressing.

Gelling agents to promote the formation of a gel, such as in gelatin desserts.

Humectants to add moisture to foods, such as shredded coconut and marshmallows, and prevent them from drying out.

Sequestrants to combine with and neutralize trace amounts of metals (such as iron and copper) in foods (such as alcoholic beverages, salad dressings and canned meats) to keep them from affecting colour, flavour and texture.

Stabilizers to prevent the settling of suspended particles, such as chocolate in chocolate milk.

Thickeners to thicken a liquid, such as in pudding or pie fillings.

6

PROCESSING AGENTS

PROCESSING AGENTS serve various purposes. They may be used to hasten natural processes (such as the bleaching and aging of flour), to prevent undesired reactions from occurring (such as the clumping of salt as a result of moisture absorption) and to initiate desired reactions (such as the curdling of milk in making cheese). In all cases, processing agents are added deliberately by food manufacturers to make processing easier and to ensure consistent results.

Processing agents include:

Acids to produce the carbon dioxide in leavening agents that makes batter bubbly and breads and other baked goods light. They are also used to give a sour or sharp flavour to some foods and to retard spoilage in butter.

Alkalies to counteract the excess acidity in foods such as chocolate and cocoa, baking powder and crackers.

Anti-caking agents to prevent powders or granules, such as table salt, from absorbing moisture, sticking together and forming clumps. They are not required to be listed on food labels.

Anti-foaming agents to prevent excessive foaming, boiling over and the formation of a film in the processing of jams and cooking oils.

Bleaching and maturing agents to hasten the natural aging and whitening process of flour and to ensure consistent results.

Clarifying agents to clear the sediment in liquids and to remove small particles and traces of minerals, such as copper and iron.

Dough conditioning agents to make dough easier to handle and reduce mixing time, resulting in better texture and volume in bakery products.

Extraction (carrier) solvents to isolate or extract desirable components, such as colours, oils and flavours, from various foods for use in other foods. They can also be used to remove caffeine from coffee beans in the decaffeination process. Residues of solvents may remain in the food. They are not required to be listed on food labels.

Firming agents to keep foods such as canned fruits and vegetables from becoming soft during heat processing and to help milk coagulate in the production of some cheeses.

Food enzymes to help initiate desired reactions, such as the formation of cheese curds.

Glazing and polishing agents to give foods such as candy and cake decorations a shiny surface or polish and/or a protective coating.

Pressure-dispensing agents to act as propellants for food packaged in aerosol cans.

Release agents to prevent food from sticking to baking surfaces during manufacture. They are not required to be listed on food labels.

What chemicals are used in decaffeinating tea and coffee? Are they of any concern?

Decaffeination involves the use of organic solvents, carbon dioxide or charcoal filters. Solvents are the most widely used decaffeination method.

Unroasted coffee is soaked in hot water containing methylene chloride, one of two decaffeination solvents permitted for coffee sold in Canada and also used for decaffeinating tea. The solvent dissolves the caffeine. The solvent is then separated and the caffeine precipitated out. The caffeine is sold to soft-drink and pharmaceutical companies. The beans are returned to the soaking water to help restore flavour and then are rinsed. Any remaining solvent residue is likely vapourized when the beans are roasted. Methylene chloride was linked to liver cancer in rats and mice in some 1982 American research, but follow-up studies have not confirmed the risk. In any case, tests have not indicated any trace of solvents of any kind in the coffee. Ethyl acetate, the other solvent approved for decaffeination processes in Canada, is generally considered harmless but it is expensive and is not used widely. Some commercial roasters who do use it have termed the process "natural decaffeination." Although ethyl acetate does occur naturally in some fruit, no decaffeination process can justly be termed as natural.

There are at least two decaffeinating methods that don't use chemical solvents at all. Liquid carbon dioxide applied at high pressure can extract caffeine from coffee beans without leaving harmful residues. The Swiss Water Process decaffeination method involves soaking the coffee, then using charcoal filters to remove the caffeine from the water. This means that no solvents are released into the environment. Some commercial roasters using solvents in decaffeination will declare that their coffee is "water-processed." Any decaffeinated coffee can be called water-processed. Look specifically for "Swiss Water-Processed" if you wish to purchase coffee that has been decaffeinated without solvents.

Adapted from *Canadian Consumer*, April 1991.

Starch-modifying agents to extend, thicken, stabilize and modify texture. They are not required to be listed on food labels.

Whipping agents to help achieve a stable whipped product.

Yeast foods to act as nutrients for yeasts in bakery products and some alcoholic beverages. Those in alcohol are not required to be listed on labels.

Other miscellaneous food additives include *coatings* (for cheeses and fresh fruits and vegetables), *plasticizing and dusting agents* (for chewing gum) and *wetting agents* (for dry beverage mixes).

7

QUESTIONABLE FOOD ADDITIVES

IF ADDITIVE SAFETY were an open-and-shut case, then all countries would ban the same chemicals. This, however, is not the case. For example, Canada permits the food colours amaranth (U.S. red dye no. 2) and carbon black, but the United States permits neither. On the other hand, Canada banned the sweetener saccharin in 1978, whereas the U.S. still allows it. These inconsistencies show that neither the science nor the politics of food additive safety is at all simple.

Listed below are those food additives permitted in Canadian food which some food additive experts suspect of having detrimental health effects on people. Some have already been proven harmful. We may see the acute effects of a food additive, particularly allergic reactions, within minutes or hours or days. (Some hyperactive children improve markedly when certain food additives are removed from their diet.) But more insidious are those effects, such as cancer, which may result from years of low-dose exposure to an additive.

For each additive, the table also lists some of the foods in or on which its use is permitted. This does not mean that all brands of a given product contain the additive. Read the label.

TABLE 3 Questionable Food Processing Additives

ADDITIVE	TYPE	POSSIBLE EFFECTS*	PERMITTED IN OR UPON
acacia gum (gum arabic)	texture-modifying agent; glazing and polishing agent	• associated with allergic reactions such as skin rash and asthmatic attacks • inadequately studied	ale; beer; cream; French dressing; light beer; malt liquor; milk; mustard pickles; porter; relishes; salad dressing; ice cream and mix; ice milk and mix; sherbet; unstandardized foods; calorie-reduced margarine; canned asparagus; canned green beans; canned wax beans; canned peas; confectionary
alum	firming agent; pH-adjusting agent; miscellaneous agent	as for potassium aluminum sulphate	same foods as for potassium aluminum sulphate
aluminum potassium sulphate	firming agent; pH-adjusting agent; miscellaneous agent	as for potassium aluminum sulphate	same foods as for potassium aluminum sulphate

ADDITIVE	TYPE	POSSIBLE EFFECTS*	PERMITTED IN OR UPON
aspartame (NutraSweet)	sweetener	• can cause brain damage and retardation among people with phenylketonuria • may cause brain chemistry changes, tumours, behaviour changes, headaches, depression, seizures, hives, vision problems, menstrual difficulties • may affect foetal brain development ◆ should be avoided by people with phenylketonuria	tabletop sweeteners; breakfast cereals; beverages; desserts; dessert mixes and toppings; dessert fillings; chewing gum; breath freshener products
benzoic acid	preservative	• associated with allergic reactions such as asthma, red eyes and skin rashes, behavioural changes, water retention, and intestinal upsets ◆ should be avoided by sensitive individuals, especially those who react to acetylsalicylic acid (ASA) and tartrazine ◆ should be avoided by people with liver problems	jam; jelly; marmalade; fruit juices; processed meat and fish; tomato pulp, puree, paste and ketchup; unstandardized foods; margarine

ADDITIVE	TYPE	POSSIBLE EFFECTS*	PERMITTED IN OR UPON
brominated vegetable oil	flavour enhancer	• accumulates in the body • linked with liver, kidney, thyroid, spleen, pancreas and testicle damage, birth defects and growth problems • may be associated with allergic reactions	citrus- or spruce-flavoured beverages
butylated hydroxy-anisole (BHA)	preservative	• accumulates in body fat • associated with kidney damage, allergies, behavioural changes, nerve and reproductive system damage, birth defects, cancer, haemorrhaging, and lung damage • reduced weight gain, liver enlargement, increased adrenal gland weights and increased blood cholesterol levels reported in lab animals • intake of large doses can lead to stomach-cramps, vomiting, dizziness and loss of consciousness, especially when consumed on an empty stomach	fats and oils; lard; monoglycerides and diglycerides; shortening; dried breakfast cereals; dehydrated potato products; chewing gum; essential oils; citrus oil flavours; dry flavours; citrus oils; partially defatted pork and beef fatty tissue; vitamin A liquids and dry vitamin D preparations to be added to food; dry beverage mixes; dry dessert and confection mixes; active dry yeast; unstandardized foods; fish; poultry meat and poultry meat by-products; margarines

ADDITIVE	TYPE	POSSIBLE EFFECTS*	PERMITTED IN OR UPON
BHA *contd.*		• pregnant mice fed BHA (or BHT) gave birth to offspring with chemical changes in the brain and abnormal behaviour patterns ◆ should be avoided by infants and young children and people allergic to ASA (Aspirin)	
butylated hydroxy-toluene (BHT)	preservative	same as for BHA	fats and oils; lard; monoglycerides and diglycerides; shortening; dried breakfast cereals; dehydrated potato products; chewing gum; essential oils; citrus oil flavours; dry flavours; citrus oils; partially defatted pork and beef fatty tissue; vitamin A liquids and dry vitamin D preparations to be added to food; parboiled rice; dry beverage mixes; dry dessert and confection mixes; active dry yeast; unstandardized foods; fish; poultry meat and poultry meat by-products; margarine

ADDITIVE	TYPE	POSSIBLE EFFECTS*	PERMITTED IN OR UPON
1,3-butylene glycol	extraction solvent	• may cause stimulation of the central nervous system leading to depression, vomiting, drowsiness, coma, respiratory failure and convulsions • renal damage may proceed to uremia and death	flavours and flavouring preparations
caffeine	flavour enhancer	• associated with high blood pressure, central nervous system damage, behavioural changes, birth defects, miscarriages, reduced fertility, heart problems, anxiety and panic attacks, insomnia, ulcer aggravation, asthmatic reactions • can cross placental barrier and affect foetus • may be associated with fibrocystic breast lumps • can cause nervousness, irregular heart-beat, noises in ears and convulsions ◆ should be avoided by pregnant and breastfeeding women and by children	cola-type beverages

ADDITIVE	TYPE	POSSIBLE EFFECTS*	PERMITTED IN OR UPON
calcium aluminum silicate	anti-caking agent	• may cause kidney damage	salt; garlic salt; onion salt
calcium carrageenan	texture-modifying agent	• in mice and rats, associated with decrease in number of live births, abnormal or underdeveloped offspring, chromosomal changes in bone marrow (rats only), inhibition of immune system and changes in blood cells • inadequate testing ◆ may best be avoided during pregnancy ◆ may best be avoided during infectious illness or disturbances of the body's metabolic processes	same foods as for carrageenan
calcium disodium EDTA	sequestering agent	• may cause skin irritations, allergic reactions, intestinal upsets, muscle cramps, kidney damage, blood in urine and mineral deficiencies	ale; beer; light beer; malt liquor; porter; stout; French dressing; mayonnaise; salad dressing; unstandardized dressings and sauces; potato salad; sandwich spread; canned shrimp, tuna, crabmeat, lobster

ADDITIVE	TYPE	POSSIBLE EFFECTS*	PERMITTED IN OR UPON
calcium disodium EDTA *cont'd*		• more information is needed on ability to cause genetic damage, birth defects and reproductive effects	and salmon; margarine; cooked, canned clams; canned ripe lima beans (butter beans); canned pinto beans
calcium furcelleran	texture-modifying agent	• inadequate testing • very similar in chemical structure to carrageenan; same concerns may apply ◆ best to avoid during pregnancy ◆ best to avoid during infectious illness or disturbances of the body's metabolic processes	same foods as for furcelleran
calcium silicate	anti-caking agent	as for calcium aluminum silicate	salt; garlic salt; onion salt; baking powder; dry cure; icing sugar; unstandardized dry mixes; meat binder; grated cheese; oil-soluble annatto
carboxy-methyl cellulose	texture-modifying agent; miscel-laneous agent	as for sodium carboxymethyl cellulose	same foods as for sodium carboxymethyl cellulose

ADDITIVE	TYPE	POSSIBLE EFFECTS*	PERMITTED IN OR UPON
carrageenan	texture-modifying agent	• may cause ulcers, particularly in infants • associated with colitis and colon cancer • associated with decreased number of live births in pregnant animals and immature skeletal structures • in animals, tends to increase permeability of blood vessels and inhibit the formation of chemicals that trigger the body's immune system	ale; beer; brawn; canned poultry; cream; French dressing; headcheese; fruit jelly with pectin; light beer; malt liquor; meat binder; meat by-product loaf; meat loaf; milk; mustard pickles; porter; potted meat; potted meat by-products; prepared fish or prepared meat; relishes; salad dressing; cottage cheese; ice cream and mix; ice milk milk and mix; evaporated milk; sherbet; infant formula; unstandardized foods; calorie-reduced margarine; sour cream; canned asparagus; canned green beans; canned wax beans; canned peas; cream cheese; processed cheese spread; cold-pack cheese
dichloromethane	carrier or extraction solvent	as for methylene chloride	same foods as for methylene chloride
disodium EDTA	sequestering agent	as for calcium disodium EDTA	dressings and sauces; sandwich spread; canned red kidney beans; canned chick peas (garbanzo beans); canned black-eyed peas; dried banana products

ADDITIVE	TYPE	POSSIBLE EFFECTS*	PERMITTED IN OR UPON
disodium guanylate	flavour enhancer	◆ should be avoided by people suffering from gout and other conditions requiring avoidance of purines	no restrictions on use
disodium inosinate	flavour enhancer	as for disodium guanylate	no restrictions on use
furcelleran	texture-modifying agent	• may cause allergic reactions • not adequately tested	ale; beer; light beer; malt liquor; porter; stout; unstandardized foods; calorie-reduced margarine; canned asparagus; canned green beans; canned wax beans; canned peas
guar gum	texture-modifying agent	• may cause allergic reactions	cream; French dressing; milk; mince meat; mustard pickles; relishes; salad dressing; cottage cheese; ice cream and mix; ice milk and mix; infant formula; sherbet; unstandardized foods; calorie-reduced margarine; sour cream; canned asparagus; canned green beans; canned wax beans; canned peas; cream cheese; processed cheese spread; cold-pack cheese

ADDITIVE	TYPE	POSSIBLE EFFECTS*	PERMITTED IN OR UPON
gum arabic	texture-modifying agent; glazing and polishing agent	as for acacia gum	same foods as for acacia gum
gum tragacanth	texture-modifying agent	as for tragacanth gum	same foods as for tragacanth gum
hydrolyzed vegetable protein (HVP)	flavour enhancer	• may cause allergic reactions • may contain up to 40% monosodium glutamate (*see* MSG) • may cause brain damage in infants	no restrictions on use
karaya gum	texture-modifying agent	• linked to allergic reactions • not adequately tested	French dressing; milk; mustard pickles; relishes; salad dressing; cottage cheese; ice cream and mix; ice milk and mix; sherbet; unstandardized foods; calorie-reduced margarine

ADDITIVE	TYPE	POSSIBLE EFFECTS*	PERMITTED IN OR UPON
magnesium aluminum silicate	dusting agent	• as for calcium aluminum silicate • talc (a form of magnesium silicate) may be contaminated with asbestos, which is suspected of causing gastro-intestinal cancer	chewing gum
magnesium silicate	anti-caking agent	• as for magnesium aluminum silicate	salt; garlic salt; onion salt; icing sugar; unstandardized dry mixes
	glazing and polishing agent; release agent		confectionery
	dusting agent		chewing gum
	coating		rice
methylene chloride	extraction solvent	• high concentrations are narcotic • can cause damage to liver, kidney and central nervous system • may cause cancer	spice extracts; natural extractives; hop extracts for use in malt liquors; green coffee beans and tea leaves for decaffeination

ADDITIVE	TYPE	POSSIBLE EFFECTS*	PERMITTED IN OR UPON
mineral oil	glazing and polishing agent	• may inhibit absorption of digestion fats • mild laxative effect	confectionery
	release agent		bakery products; confectionery; seeded raisins
	coating		fresh fruits and vegetables
	lubricant		sausage casings
	binding agent and coating		salt substitutes
mono-ammonium glutamate	flavour enhancer	• associated with brain and learning disorders, visual and behavioural problems and allergic reactions ◆ should be avoided by pregnant women and young children	no restrictions on use
mono-potassium glutamate	flavour enhancer	as for monoammonium glutamate	no restrictions on use

ADDITIVE	TYPE	POSSIBLE EFFECTS*	PERMITTED IN OR UPON
monosodium glutamate (MSG)	flavour enhancer	• may cause allergic reactions • causes brain and eye damage to lab animals • may not be safe for human foetuses or infants • linked to reproductive disorders, fertility problems, learning disorders, depression, irritability, other mood changes • reactions can mimic symptoms of heart attack ◆ should be avoided by nursing mothers and infants	no restrictions on use
nitrate	preservative	• can be converted to nitrites by digestive juices	see specific nitrate (potassium nitrate; sodium nitrate)
nitrite	preservative	• can be toxic in moderate amounts • linked to cancer, brain damage, arthritis and anemia • impairs the blood's ability to carry oxygen which, in extreme cases, can lead to death	see specific nitrite (potassium nitrite; sodium nitrite)

ADDITIVE	TYPE	POSSIBLE EFFECTS*	PERMITTED IN OR UPON
nitrite *cont'd*		• may be converted to nitrosamines which are known to cause cancer ◆ should be avoided during pregnancy ◆ should be avoided by infants ◆ should be avoided by cancer patients ◆ should be avoided by people with reduced stomach acidity (e.g. those with ulcers)	
potassium aluminum sulphate	firming agent; pH-adjusting agent; miscellaneous agent	• inhibits retention of phosphorous • aluminum component can aggravate kidney disease and even cause kidney damage ◆ anyone regularly taking antacid medication containing aluminum should make certain of sufficient phosphorous in the diet to counteract effects of aluminum	pickles and relishes; unstandardized foods; flour; whole wheat flour; ale; baking powder; beer; light beer; malt; practice liquor; oil-soluble annatto; porter; stout
potassium bisulphite	preservative	• can cause allergic reactions, asthma and anaphylactic shock in small doses • can destroy vitamin B_1 (thiamine)	same foods as for sulphurous acid

ADDITIVE	TYPE	POSSIBLE EFFECTS*	PERMITTED IN OR UPON
potassium bisulphite *cont'd*		◆ should be avoided by asthmatics; uncertain safety for others	
potassium bromate	bleaching, maturing and dough-conditioning agent	• linked to tumours, kidney failure, cancer and mutations • may be linked to central nervous system damage • promotes oxidation of fats and oils	flour; whole wheat flour; bread; unstandardized bakery foods
potassium metabi-sulphite	preservative	as for potassium bisulphite	same foods as for sulphurous acid
potassium nitrate	preservative	as for potassium nitrite	dry sausage; semi-dry sausage; preserved meat and preserved meat by-products; ripened cheese
potassium nitrite	preservative	• converts to nitrosamines, powerful cancer-causing agents	preserved meat and preserved meat by-products; preserved poultry meat and preserved poultry meat by-products
propyl gallate	preservative	• linked to liver and kidney damage, cancer, lymphoma, reproductive problems and allergic reactions	fats and oils; lard; monoglycerides and diglycerides; shortening; dried breakfast cereals; dehydrated potato products;

ADDITIVE	TYPE	POSSIBLE EFFECTS*	PERMITTED IN OR UPON
propyl gallate *cont'd*		• can cause stomach or skin irritation especially in people who suffer from asthma or are sensitive to ASA (Aspirin)	chewing gum; essential oils; dry flavours; citrus oils; unstandardized foods; dried cooked poultry meat; margarine
1,2-propylene glycol (1,2-proanediol)	anti-caking agent	• large oral doses in animals have been reported to cause central nervous system depression and slight kidney changes • may cause allergic rash	salt
	extraction solvent		oil-soluble annatto; annatto butter colour; annatto margarine colour
	humectant		unstandardized foods
shellac	glazing or polishing agent	• no studies on the biological effects in humans or in experimental animals from ingestion	cake decorations; confectionery
sodium aluminum silicate	anti-caking agent	as for calcium aluminum silicate	salt; icing sugar; dried whole egg, egg white, egg yolk, whole egg mix, egg yolk mix; garlic salt; onion salt; unstandarized dry mixes
sodium benzoate	preservative	as for benzoic acid	same foods as for benzoic acid

ADDITIVE	TYPE	POSSIBLE EFFECTS*	PERMITTED IN OR UPON
sodium bisulphite	preservative	as for potassium bisulphate	same foods as for sulphurous acid
sodium carboxy-methyl cellulose	texture-modifying agent; miscellaneous agent	• may have caused cancer when injected under the skin of rats • shown to cause cancer in animals when ingested	cream; French dressing; milk; mustard pickles; relishes; salad dressing; cottage cheese; ice cream and mix; ice milk and mix; sherbet; unstandardized foods; glaze of frozen fish; processed cheese; cream cheese; processed cheese spread; cold-pack cheese
sodium dithionite	preservative	as for potassium bisulphate	same foods as for sulphurous acid
sodium meta-bisulphite	preservative	as for potassium bisulphite	same foods as for sulphurous acid
sodium nitrate	preservative	as for potassium nitrite	sausage; preserved meat and meat by-products; ripened cheese
sodium nitrite	preservative	as for potassium nitrite	preserved meat and meat by-products; side bacon; preserved poultry meat and poultry meat by-products

ADDITIVE	TYPE	POSSIBLE EFFECTS*	PERMITTED IN OR UPON
sodium silicate	miscellaneous agent	as for calcium aluminum silicate	canned drinking water
sodium sulphite	bleaching, maturing and dough-conditioning agent	as for potassium bisulphite	biscuit dough
	conditioning agent		canned flaked tuna
	preservative		same foods as for sulphurous acid
sulphurous acid	preservative	as for potassium bisulphite	cider; honey wine; wine; ale; beer; light beer; malt liquor; porter; stout; jam and jelly; molasses; some marmalade; frozen sliced apples; fruit juices; gelatin; mince-meat; pickles and relishes; syrup; tomato ketchup, paste, pulp and puree; beverages; dried fruits and vegetables; unstandarized foods; frozen mushrooms; dextrose anhy-drous; dextrose monohydrate; glucose or

ADDITIVE	TYPE	POSSIBLE EFFECTS*	PERMITTED IN OR UPON
sulphurous acid *cont'd*			glucose syrup; crustaceans; grapes; sliced potatoes for restaurants
	pH-adjusting agent		gelatin
tragacanth gum	texture-modifying agent	• associated with allergic reactions • not adequately tested	French dressing; mustard pickles; salad dressing; relishes; cottage cheese; ice cream and mix; ice milk and mix; sherbet; lumpfish caviar; unstandardized foods; calorie-reduced margarine; cream cheese; cold-pack cheese; processed cheese spread
xylitol	sweetener	• possible cancer-causing agent • can cause stomach upsets in large amounts • may help prevent tooth decay	chewing gum

* The information on possible effects of food additives has been excerpted from the following sources:
Freydberg, Nicholas and Willis Gortner. *The Food Additives Book*. Mount Vernon, New York: Consumers Union, 1982.
Goodman, Robert. A *Quick Guide to Food Additives*. 2nd Ed. San Diego: Silvercat Publications, 1990.
Harte, John, Cheryl Holdren, Richard Schneider and Christine Shirley. *Toxics A to Z*. Berkeley: University of California Press, 1991.
Winter, Ruth. *The Consumer's Dictionary of Food Additives*. New York: Crown Publishers, 1989.

It should be noted that not all of the same reported effects were mentioned in every text referenced above. The effects as stated in the table should not be taken to reflect the information as presented in each text, but, rather, understood to be a compilation of the effects reported in all texts combined.

This list represents about 20 per cent of the approximately 350 food additives permitted in Canada (not including the 1,500 flavouring agents and enhancers). It should be noted that roughly the same number of additives has been removed from use in Canadian food since 1964, when our food laws assumed their present form. Over the same period, a similar number of new additives have found their way into the food supply.

Listing these questionable additives brings up a number of questions:

- Can a combination of additives have a greater effect than the sum of the effects of individual additives? In other words, what about the synergy (chemical interaction producing heightened effects) of food additives with each other and with food ingredients?
- Can smoking aggravate the effects of some food additives, similar to the way that it aggravates the health risks of, for instance, oral contraceptives and asbestos dust?
- What proportion of behavioural effects can be attributed to food additives? Laboratory tests on animals check primarily for physical effects. And even certain physical effects, such as nausea, headaches and joint pain, cannot be easily observed.

CONCERNS OR QUESTIONS?

If you have concerns or questions about a particular additive, especially if you want to know about *all* foods in which a particular additive might be found, contact the nearest Health Protection Branch of Health and Welfare Canada. You will find them listed in the Government of Canada Blue Pages of your telephone directory.

PART TWO:

UNINTENTIONAL ADDITIVES and Other Issues

Unintentional or accidental additives are additives that are intentionally used on or in food when it is grown, harvested, shipped and processed, but which may unintentionally remain in or on the food when we eat it. Unintentional additives may be added at any stage of the food production cycle.

STRICTLY SPEAKING pesticide residues, hormones and antibiotics in animal products and food-packaging chemicals are not considered food additives by the Canadian government. However, since they may remain in and affect the safety of the food you eat, they are included here under the heading of unintentional additives. Although these substances are deliberately, or intentionally, used, they are not necessarily meant to remain in or on the food when it is eaten.

Up until 1989, food irradiation was considered a food additive in the *Food and Drug Regulations*. Although it is now designated as a process, food irradiation is included here as an issue of concern as it can alter foods in undesirable ways. It is important that consumers be aware of and understand the issues surrounding food irradiation to make informed decisions when purchasing foods and even to affect government legislation regarding its use.

Finally, labelling is included in this section to help consumers to be better informed and protected when comparing different food products and making choices in their purchases.

8

PESTICIDES AND COATINGS

PESTICIDES ARE chemicals designed to kill organisms that are regarded as pests. (Unfortunately they often also kill a lot of "non-target" plants and animals.) They are used in agriculture, forestry, industry and the home and garden. In Canada, by far the greatest volume – 86 per cent – of pesticides is used in agriculture.

Pesticides are sprayed or dusted on and around seeds and plants to protect them from disease and to control damage caused by insects, worms, rodents, and fungi. Pesticides are also used to improve the appearance of fresh produce and protect it from bruising and spoilage. They are applied to the food that humans eat directly, to the grains and grasses that meat-producing animals eat and to farm animals to combat parasites.

This means that consumers may be exposed to pesticide residues in two ways. We may encounter residues directly on and in the fruits, vegetables and grains that we eat. We may also ingest them second-hand when we eat animal products. In animals fed pesticide-treated food or sprayed directly with pesticides, residues accumulate in their fatty tissues. When these animals are eaten by other animals, the residues of the prey animals are then passed on to and accumulate in the fatty tissues of the predatory animals. When we eat fatty tissue (meat) we ingest the pesticide residues that have accumulated there from

CANADIAN AGRICULTURE IS LOW-CHEM

In fact, Canadian agriculture is almost no-chem in comparison to other major crop-growing countries. An environmental report issued by the Paris-based Organization for Economic Co-operation and Development confirms that Canadian farmers use chemical fertilizers and pesticides sparingly. On average, an acre of arable land gets seven times more nitrogenous fertilizer in Germany or the United Kingdom than in Canada, and five or six times more pesticide. An acre in Japan gets eighteen times more pesticide and five times more nitrogenous fertilizer. Even in the United States, each arable acre on average gets twice the pesticide and almost twice the nitrogenous fertilizer of a Canadian acre.

Source: "At Press Time," *Country Guide*. July 1991, p.10.

all the foods that the animal has previously eaten. The residues then accumulate in our own fatty tissues.

Many people are concerned about the effects of eating such residues, especially over a lifetime. As with most other safety issues, the questions surrounding pesticides and pesticide residues are complicated and often controversial. For as many experts as there are who say that pesticide residues in our food are safe, there is an equal number who disagree.

GOVERNMENT REGULATION – A SHARED RESPONSIBILITY

All pesticides must go through a standard approval process before they are allowed to be used in Canada. (This process is currently undergoing review and revision.) Under the Pest Control Products Act, Agriculture Canada is responsible for registering pesticides. Applicants wishing to have a new pesticide registered perform various tests on the chemical and then submit the data to Agriculture Canada, which reviews them to determine what risks, if any, may be associated with the use of the pesticide. They decide if and how the product is effective, how it may be used, what directions are needed for using it and all conditions for selling it. Agriculture Canada primarily considers the efficacy of the pesticide. Environment Canada is also part of the pesticide registration process.

INTEGRATED PEST MANAGEMENT (IPM)

Agriculture Canada and various provincial ministries are helping farmers to cut back on the amount of pesticides they use and are promoting an alternative method of pest control called Integrated Pest Management (IPM). Instead of spraying chemicals at routine intervals, crops are visually checked once or twice a week for signs of insects or plant disease. When a problem is identified, an IPM consultant then makes a recommendation on how to deal with it. If insects are just beginning to harm the crop, the consultant may recommend that pesticides be applied to only the outside rows. Other methods include introducing harmless bugs to feed on those doing the damage and harvesting the crop early, before the damage becomes too great.

With these alternative methods, the use of pesticides may be greatly reduced – often by 50 per cent. Risks to the farmers are also reduced by spraying less often. For the consumer, fewer pesticides sprayed means fewer residues on or in foods.

IPM is used most in British Columbia, Ontario and Quebec and to a lesser extent elsewhere. Unfortunately, produce currently does not carry any label to indicate that they have been grown using IPM methods. Much of the local produce in British Columbia has been grown this way. There, 85 per cent of carrots and onions and almost all head lettuce and greenhouse crops are grown using IPM – almost entirely without pesticides.

It should be noted that, although it is definitely a step in the right direction, IPM has been heavily criticized by advocates of totally organic agriculture. IPM still allows the use of pesticides. Methods that avoid toxic pesticides entirely are preferable to IPM.

Health and Welfare Canada, and specifically the Health Protection Branch (HPB), is responsible for assessing the risks of pesticide residues that reach the consumer through food consumption. Before Agriculture Canada registers a pesticide for use on food, the HPB does its own evaluation.

The HPB requires that manufacturers submit the following information to them:

- specifications and composition of the substance to be used as a pesticide
- the physical and chemical properties of the substance
- plant and animal metabolism studies
- evidence that the product is effective and practical
- the amount to be applied, frequency and time of application
- satisfactory method of analysis for determining residues in foods
- toxicity studies
- a proposed maximum residue limit for each food

After these data have been submitted, there are three steps in the process of determining the maximum residue limit for a particular pesticide:

Toxicity Studies: The manufacturer must perform several types of toxicity studies. These studies involve looking at the long- and short-term effects the chemical has when ingested, applied to skin or inhaled. These tests are performed on animals and help to determine how harmful the chemical might be for humans.

Acceptable Daily Intake: On the basis of these studies, and any other data that are available, the HPB decides what quantity of the chemical is safe for humans to consume over their lifetime. This level is set by finding the lowest dose (expressed in milligrams per kilogram of body weight) that produced no observable adverse effects in the test animals. This "no-effect" level is then divided by a safety factor, usually 100. This amount is called the "Acceptable Daily Intake" or ADI. According to John Salminen of Health and Welfare Canada, "The acceptable daily intake is the quantity of the substance that humans can consume on a daily basis, for a lifetime, with reasonable assurance that their health will not be threatened."

The purpose of the safety factor is to take into account the possibility that humans are more sensitive to the chemical than are the test animals. As well, some humans, especially children, pregnant women and older people, may be more sensitive than others.

Generally, Canadian ADI levels are the same as, or more stringent than, those recommended by the WHO.

Setting Maximum Residue Limits: The HPB determines how much of a pesticide is likely to remain on or in foods at the time they are sold. In calculating this they assume that the pesticide has been used properly. Considering all the foods to which the pesticide might be applied, the HPB works out how much an average person is likely to consume in a day. If the amount of residue that is likely to remain on or in foods and be consumed by a person in a day, from all food sources, does not exceed the ADI, then that residue level is accepted as the maximum residue limit.

The amount of a pesticide residue that is allowed to remain depends upon how toxic the chemical is. For very toxic pesticides, no detectable residues are allowed to remain. Some pesticides are not likely to leave a residue because of their chemical structure or because of the way they are applied. For these chemicals no maximum level is set. The same is true for those pesticides that have a low toxicity.

There are approximately 250 to 300 chemicals registered in Canada for use in the production and handling of foods. These chemicals are combined into thousands of different formulations. Of the base chemicals, the HPB has established residue limits for about 100. Residue limits are usually expressed in parts per million (ppm) or parts per billion (ppb). If a specific maximum residue limit is not set, then the limit is 0.1 ppm.

EBDCs

- EBDCs (ethylene bisdithiocarbamates) are a group of chemically related fungicides registered in Canada for use on vegetables and fruits (such as potatoes, tomatoes, apples, grapes). The chemicals are Ferbam, Mancozeb, Maneb, Metiram, Nabam, Zineb and Ziram.
- The main concern is not EBDCs themselves, but their breakdown product known as ethylene thiourea (ETU). ETU can form during processing or cooking. In high doses it has been shown to cause cancer and birth defects in laboratory animals.
- Residues of these fungicides do not penetrate foods. They remain on the surface and can be removed by thorough washing.
- In the United States there has been a lot of concern over EBDC fungicides. According to the Health Protection Branch of Health and Welfare Canada, the U.S. situation is not applicable to Canada because of regulatory action taken here since the mid-seventies.
- EBDCs and ETU were re-evaluated in Canada in the mid-seventies. Registrations were cancelled for use on thirty crops and allowable maximum residue limits were reduced for other crops. Maximum residue limits currently range from 0.1 to 7.0 parts per million.

Adapted from: *Issues: Background Information on Ethylene Bisdithiocarbamate (EBDC) Fungicides*. Health and Welfare Canada, Health Protection Branch. September 12, 1989.

But pesticides themselves aren't the only concern. Pesticides may break down into other products during food processing. These chemicals may be more or less toxic than the original pesticide. If the HPB finds these breakdown products, known as metabolites, in the bodies of test animals they too are tested. If they are significantly more or less toxic than the pesticide itself, separate residue limits may be set for them.

If a pesticide is approved by both Agriculture Canada and Health and Welfare Canada it is registered for use. This process may take up to ten years to complete and cost millions of dollars. It is then up to provincial ministries to assist in the control, distribution and handling of pesticides in their provinces. Basically, the federal government regulates registration of the chemical while provincial governments regulate the use of the chemical. Provincial authorities may set more stringent regulations than those legislated by the federal authorities, but they may not make them less stringent.

WHY WORRY?

Why are so many people worried about pesticides? Given all of this extensive testing and apparently strict regulatory process it would seem that all the pesticides in use in Canada must be safe. Unfortunately this isn't the case. All pesticides are potentially toxic to humans. Of course, *anything* can be toxic if it is present in large enough amounts. And it is the dose that makes the difference. It depends, in part, on how much you consume and how often and also on how sensitive you are to the chemical in question.

Effects on Children

There are specific concerns about the adverse effects that these chemicals may have on children. Acceptable daily intake levels are determined according to an adult body's metabolism. There is some suggestion that, because children's bodies are still developing, toxins may be deposited there more readily. Since their intestinal tracts are easily penetrated, children may be more vulnerable to carcinogens and other toxins. They may also be more likely to experience genetic damage. And children's eating patterns tend to be different from those of adults. For young children especially, fruit and fruit juices tend to make up a larger part of their diet than they do in adults'. According to John Salminen of HWC, "the potential exposure of children to such chemical residues is taken into consideration during the review process, both in the evaluation of the results of toxicological studies and the assessment of potential intake resulting from residue levels in foods which are most likely to be consumed in greater quantities by infants or children." Even with the built-in safety factor, some people still fear that acceptable intake levels do not adequately consider the differences in the eating patterns of children and adults.

Combined Effects?

Those who question the safety of pesticides also question the government's practice of basing everything on toxicity studies. When a chemical is studied in this way, scientists are able to tell only how toxic that one chemical is on its own. They haven't seriously investigated the interactions that may occur between the hundreds or even thousands of chemicals that we encounter every day. How do particular pesticides present on tomatoes interact with the preservatives that are used in making those tomatoes into spaghetti sauce? What effects might there be from consuming a pesticide and breathing in exhaust from cars or fumes from paint? These combined effects may be impossible to determine or predict. And there is also the question of what happens as these chemicals accumulate in our bodies over a lifetime.

Comparatively Speaking

Of course, many experts tell us that the risks we are exposed to from pesticides are small compared to the risks we encounter from, say, food poisoning. It is true that there are many more instances of death and illness directly linked to food poisoning than there are to the presence of pesticides in our food supply.

But the indirect (cumulative and synergistic) sub-lethal effects of pesticides have not been adequately studied. Besides, the presence of one risk does not justify the presence of another.

Banned Pesticides

Toxicity testing using lab animals usually involves administering massive doses of a chemical over relatively short periods of time and then extrapolating the resultant health effects to lower doses over longer time. This may miss serious problems related to chronic exposure. In addition, humans may react differently than the test species used. These problems may only come to light once the chemical goes into widespread use.

If, as the government assures us, all pesticides undergo rigorous testing, then why are some of them later deemed to be unsafe and then banned? Because scientists can base their conclusions only on the data they have available to them. Methods for determining the toxicity of chemicals have become more refined and precise than they were, say, thirty years ago. This means that researchers are continually becoming better able to determine the danger, or safety, of a particular chemical. With these more accurate methods, some pesticides earlier thought to be safe have now been deemed to pose a hazard. It is possible, then, that many of the pesticides that are currently considered to be safe may, in a few years, with even more precise methods of measurement, be determined to pose health hazards and also be banned.

Some pesticides that have already been banned in Canada, such as DDT, continue to show up in our food supply. They may be used in other countries and be present on or in imported foods. And some of the most toxic pesticides tend to persist in the environment. They may remain in the soil, air or water for many years, crossing territorial boundaries, after they are no longer in active use.

More than Just Consumer Health

Aside from the negative health effects that pesticides may have on consumers, we must also consider the health of farmers, animals and the environment:

- Workers who apply pesticides may face the greatest risks. Pesticide exposure can cause cancer, birth defects, nervous disorders, and other serious diseases. In developing countries especially, workers who are illiterate or who do not speak the language in which the pesticide's label is printed may not understand the directions for application. They may not properly protect themselves or be provided with the means to protect themselves when spraying. The WHO estimates that pesticide use throughout the world may be responsible for the deaths each year of 20,000 field workers and farmers. Up to one million people suffer mild to severe side effects annually.
- Birds, bats and bees seem to be more sensitive to some pesticides than are humans. Birds and bats feed on crops and insects that have been sprayed. Bees may be sprayed inadvertently. In both cases the effects can be deadly for these animals. Having fewer birds, bees or any plant or animal species affects all parts of the ecosystem. For example, many birds and bats eat mosquitoes. If a large number of these animals are killed there will not be enough left to keep the population of mosquitoes down. If too many honeybees are killed there will not be enough to pollinate food crops. These are just a couple of the almost endless number of potential consequences.
- Forty years ago farmers lost one third of their crops to damage by pests. Today farmers use three times as many pesticides as they did then. And farmers still lose about

one-third of their crops to damage by pests. This is because some pest species have become resistant to particular pesticides. When the crop or ground is sprayed with insecticide, for example, the weak insects are killed, but stronger ones may survive and continue to breed. The next time the insecticide is applied, again the weaker insects die and the strong ones live on and multiply. As this goes on and on, eventually all that is left is a population of strong insects completely resistant to the pesticide. The chemical is no longer effective for that pest species and a new chemical must be devised. With the new chemical, the process repeats itself.

- Of the pesticides that are used, less than 0.1 per cent actually stays where it is applied. The remaining 99.9 per cent can contaminate both surface water and ground water, air, other soil and other food.
- When we rely on pesticides we tend to forget to ask why insect or plant "pests" flourish at all. We concentrate on using an end-of-the-line cure rather than considering preventative measures.

RESIDUES

Even if a pesticide is determined not to pose a significant hazard at a given level, how can we be assured that this level is not exceeded (even if accidentally) by farmers in its application? Programs to monitor actual pesticide residues on or in food products are carried out by both Agriculture Canada and Health and Welfare Canada. Some provinces also test for pesticide residues.

Within the Health Protection Branch, the Field Operations Directorate analyzes approximately 2,000 samples of imported and domestic foods each year. An investigation is conducted

WHO IS GROWING ORGANIC NEAR YOU?

The Canadian Organic Growers offers a *Directory of Organic Agriculture*, which lists more than 1,200 associations, growers, wholesalers, retailers and suppliers of organic foods, services and information. Also listed are organizations involved in research and education, and groups that focus on environmental and health concerns. The listings are organized according to province and the directory also includes contacts in the United States and Europe. It is an invaluable (and inexpensive) sourcebook for those looking to buy or sell organic foods.

A copy of the directory may be ordered from: COG Directory, Box 6408, Station J, Ottawa, ON K2A 3Y6

whenever excessive residue levels are detected. About half of the samples taken are of imported foods. These are expected to contain more residues than domestic foods since pesticides may be applied during shipping and storage.

Agriculture Canada also has a pesticide residue monitoring program. The Agri-Food Safety Division tests for more than 190 pesticides and their breakdown chemicals. But not all foods are tested. Some food items are given higher priority for testing because of the toxicity of the pesticides likely to be applied to them and because we consume these foods in large amounts. Foods of high testing priority include tomatoes, beef, veal, potatoes, oranges, grapefruit, nuts, apples, pork, fats and oils, lettuce, chicken, turkey, carrots, grapes, beans and corn. Both raw and processed forms of these foods are tested. Of processed foods, canned tomato paste is the most closely monitored. Permethrin, an insecticide registered for use on tomatoes in Canada and Florida, has been shown to cause malignant tumours at high doses in laboratory animals. According to *Canadian Consumer* magazine (1989, No. 7), tests in the United States revealed that, in certain circumstances, permethrin can concentrate by 230-fold in canned tomato paste. This means that it may exceed the 0.5 parts per million (ppm) maximum residue limit placed on it. Since this discovery, the U.S. Food and Drug Administration has ruled that Florida tomatoes may only be sold fresh, they are not allowed to be processed commercially. Canada has no such regulation.

ARE WE PROTECTED?

Once again the government is claiming that it is adequately protecting our health. And once again some people are not quite comforted by these efforts. Here are some of their concerns:

- Health and Welfare Canada and Agriculture Canada test less than 1 per cent of all imported food and domestic food for pesticide residues.
- Not all of the residues that may be present can actually be

ARE THE DATA ACCURATE?

During the mid-1970s, Industrial BioTest Laboratories submitted safety data on hundreds of chemicals on behalf of their manufacturers. One-quarter of pesticides licensed for use in Canada were licensed on the basis of these data. It was later discovered that many of the tests had been improperly conducted. The chemicals had to be retested and re-evaluated by Health and Welfare Canada for potential health risks. Some pesticides were banned as a result of this re-evaluation. Agriculture Canada is still conducting tests to determine the effects these pesticides have on wildlife and the environment. Many of these are still registered and in use.

WHAT YOU CAN DO ABOUT . . . PESTICIDES

- Wash all produce in a mild solution of regular dish detergent and water, water alone or a dilute mixture of water and vinegar. (Special cleaners are unnecessary.) Be sure to rinse thoroughly. Whenever possible, also peel fresh fruits and vegetables. You will lose some of the vitamins by doing so, but you'll gain in the assurance that your food is safer.
- Remember that some pesticides end up in the food. These cannot be removed simply by peeling or washing the outside layer. Consider buying organic foods to minimize exposure to these pesticides.
- Pesticide residues, and most other toxic chemicals, tend to concentrate in the fat of animals. Reducing your consumption of animal products will reduce your pesticide exposure. If you do eat animal products, trim fat from meat and poultry, remove the skin from poultry and fish, and avoid using drippings in gravies or broths. Choose low-fat dairy products such as skim milk and low-fat cheeses.
- Choose locally grown produce over imported varieties. Produce grown locally doesn't require as many pesticides as produce shipped a long distance and stored for an extended period. Buy directly from the grower when possible to ensure freshness. In the grocery store, check the signs to see where the produce was grown. (But be careful. The label "Canada No. 1" does not mean that the food was grown in Canada. It is a grading label and only means that the item has met some standard of quality.) If the origin is not indicated, speak to the produce manager.
- If you cannot find organic foods in your grocery store, let the manager know you'd like to see them stocked. Stores will only stock products if they are assured there is a demand for them.
- Consider growing some of your own food. You don't need a huge garden to grow things like tomatoes or lettuce. And herbs can be grown in small pots on your window sill. (It is a good idea to have your soil tested before planting to see whether it is free of toxins.)

detected. However, of those that can be detected, tremendous progress has been made in recent years in the ability to detect their presence at extremely low levels.

- When a sample of a shipment is taken for testing it may take weeks before the lab results come in. In this time the product may well have made it to grocery store shelves and possibly into the hands (and mouths) of consumers.

All of this talk can be summarized into two general concerns: Are the levels of pesticide residues that the government considers to be "safe" actually safe? Does the government adequately test for pesticide residues? And what about the residues that can't be detected?

In consideration of personal health (not to mention the health of farmers and other workers who handle the chemicals, the health of animals and the health of the environment) it would be prudent to avoid foods treated with synthetic pesticides whenever possible.

RESIDUES IN IMPORTED PRODUCE

Imported produce may have residues of pesticides that are banned in Canada, in addition to those that are allowed. It is interesting to note that the *Food and Drug Regulations* list maximum residue limits for residues of some pesticides that have been banned in Canada. Some tolerances are allowed so that we can continue to import produce. That is, we won't allow the pesticide to be used here in Canada, but we are willing to import food it has been used on.

The produce that comes from areas such as Southeast Asia and Eastern Europe is given special attention since little is known about food safety practices there. Produce from areas such as Central and South America are also carefully checked because they are known to use a wide variety of pesticides and sometimes in large amounts.

TABLE 4 What Pesticides May Be Used?

The following chart lists pesticides that may be used on selected food items sold in Canada.

FOOD	AGRICULTURAL CHEMICAL	
apples	azinphos-methyl benomyl,carbendazim and thiophanate-methyl bromophos captan carbaryl chlorobenzilate cypermethrin daminozide diazinon dichlone dicofol dimethoate diphenylamine dodine endosulfan ethephon ethion ethoxyquin fenbutatin oxide ferbam folpet formetanate hydrochloride	lindane malathion methidathion methomyl methoxychlor mevinphos monocrotophos nicotine parathion permethrin phosalone phosmet piperonyl butoxide pirimicarb propargite pyrethrins sodium orthophenyl phenate tetrachlorvinphos tetradifon thiabendazole thiram ziram
bananas	carbaryl thiabendazole	thiram
beans	azinphos-methyl benomyl, carbendazim and thiophanate-methyl carbaryl chlorthal-dimethyl	cypermethrin diazinon dicofol dimethoate disulfoton endosulfan

FOOD	AGRICULTURAL CHEMICAL	
beans *cont'd*	ethion	parathion
	ferbam	permethrin
	iprodione	piperonyl butoxide
	malathion	propargite
	methoxychlor	pyrethrins
	naled	sethoxydim
	nicotine	ziram
butter	aldrin and dieldrin	dicofol
	benomyl, carbendazim and thiophanate-methyl	endosulfan
		endrin
		heptachlor
	chlordane	lindane
	DDT	
cabbage	azinphos-methyl	methoxychlor
	carbaryl	naled
	chlorothalonil	nicotine
	chlorthal-dimethyl	parathion
	cypermethrin	permethrin
	diazinon	ziram
	disulfoton	dimethoate
	endosulfan	EBDCs (ethylenebisdithiocarbamate
	ferbam	
	lindane	methomyl
	malathion	mevinphos
	methamidophos	
carrots	benomyl, carbendazim and thiophanate-methyl	ferbam
		malathion
		maleic hydrazide
	carbaryl	methoxychlor
	carbofuran phenolic metabolites	parathion
		sodium orthophenyl phenate
	carbofuran	
	chlorothalonil	trifluralin
	diazinon	ziram
	dichloran	
cauliflower	azinphos-methyl	carbaryl

FOOD	AGRICULTURAL CHEMICAL	
cauliflower *cont'd*	chlorothalonil	lindane
	chlorthal-dimethyl	malathion
	cypermethrin	methamidophos
	diazinon	methoxychlor
	dimethoate	mevinphos
	disulfoton	naled
	EBDCs	nicotine
	endosulfan	parathion
	ferbam	ziram
cheese	aldrin and dieldrin	dicofol
	benomyl, carben-dazim and thio-phanate-methyl	endosulfan
		endrin
		heptachlor
	chlordane	lindane
	DDT	
citrus fruits	2,4-D	folpet
	azinphos-methyl	formetanate hydro-chloride
	benomyl, carben-dazim and thio-phanate-methyl	imazalil
		methidathion
	biphenyl	methomyl
	carbaryl	mevinphos
	chlorobenzilate	naled
	cypermethrin	parathion
	diazinon	phosalone
	dicofol	propargite
	dimethoate	sodium orthophenyl phenate
	diuron	
	ethephon	tetradifon
	ethion	thiabendazole
	fenbutatin oxide	
corn	alachlor	hydrogen cyanide
	carbaryl	methoxychlor
	diuron	nicotine
	ferbam	parathion
dry beans	alachlor	

FOOD	AGRICULTURAL CHEMICAL	
eggs	DDT	fungicides
	EBDCs	ethoxyquin
fish	DDT	
fresh vegetables	DDT	
garlic	anilazine	folpet
	chlorthal-dimethyl	malathion
	dichloran	
grapes	azinphos-methyl	folpet
	benomyl, carben-	iprodione
	dazim and thio-	lindane
	phanate-methyl	malathion
	captan	methidathion
	carbaryl	methomyl
	cypermethrin	methoxychlor
	daminozide	parathion
	diazinon	permethrin
	dichloran	phosalone
	dicofol	phosmet
	diuron	piperonyl butoxide
	EBDCs	propargite
	endosulfan	pyrethrins
	ethephon	tetrachlorvinphos
	ethion	tetradifon
	ferbam	ziram
lettuce	carbaryl	malathion
	chlorthal-dimethyl	methamidophos
	diazinon	methomyl
	dichloran	methoxychlor
	dimethoate	mevinphos
	disulfoton	naled
	EBDCs	nicotine
	endosulfan	parathion
	ferbam	permethrin
	folpet	ziram
	lindane	

FOOD	AGRICULTURAL CHEMICAL	
meat and meat by-products	alachlor aldrin and dieldrin BHC isomers, except lindane carbaryl chlordane chlorpyrifos coumaphos DDT	dicofol endosulfan ethion ethoxyquin heptachlor lindane methoxychlor tetrachlorvinphos
milk	alachlor aldrin and dieldrin BHC isomers, except lindane chlordane DDT	dicofol endusulfan endrin flucythrinate heptachlor lindane
onions	anilazine azinphos-methyl carbofuran phenolic metabolites carbofuran chlorthal-dimethyl diazinon dichloran ferbam	folpet lindane malathion maleic hydrazide nicotine parathion sethoxydim ziram
potatoes	aldicarb anilazine carbaryl carbofuran carbofuran phenolic metabolites chlorpropham	chlorthal-dimethyl disulfoton diuron malathion maleic hydrazide metribuzin thiabendazole
raw cereals	malathion piperonyl butoxide	pyrethrins
rice	hydrogen cyanide methoxychlor	naled

FOOD	AGRICULTURAL CHEMICAL	
soybeans	alachlor chlorthal-dimethyl dicofol fluazifop-butyl	methoxychlor naled sethoxydim
tomatoes	anilazine azinphos-methyl benomyl, carben-dazim and thio-phanate-methyl captan carbaryl chlorothalonil chlorthal-dimethyl cypermethrin daminozide diazinon dichloran dichlorvos dicofol dimethoate disulfoton EBDCs endosulfan ethephon ethion fenbutatin oxide	ferbam folpet iprodione lindane malathion methamidophos methoxychlor mevinphos monocrotophos naled nicotine parathion permethrin piperonyl butoxide pyrethrins sethoxydim sodium orthophenyl phenate tetradifon thiram ziram
wheat	carbaryl chlormequat clopyralid diuron	ethephon hydrogen cyanide methoxychlor

Source: Extracted from "Maximum Residue Limits for Agricultural Chemicals," Division 15, Table II, *Food and Drug Regulations* (Part B, Foods).

Note: Aldrin, deildrin, thiophanate-methyl, chlordane, DDT, endrin, ethion, heptachlor and toxaphene are used only to a limited extent in Canada. However, they may be present in foods imported from other countries.

Agriculture Canada tested a total of 3,953 shipments of imported fresh and processed fruits and vegetables in 1990-91. Only pesticide residues that can be detected are listed in their report. Of these, 82 shipments, or 2 per cent, exceeded Canadian maximum residue limits.

REMOVAL OF PESTICIDES FROM FRESH PRODUCE

Now you know which pesticides are allowed to be used on some fruits and vegetables. Despite the fact that in most cases pesticide residues are found to be within acceptable Canadian limits, you may want to exercise some caution and avoid as many of the residues as possible in your diet. The following table outlines various common pesticides and indicates whether residues can be removed by washing or peeling fruits and vegetables, or whether the residues penetrate the whole food and cannot be removed.

TABLE 5 Can the Residue Be Removed?

PESTICIDE	REMOVAL OF RESIDUES
Acephate	Residues permeate the produce and probably cannot be removed with washing. Cooking or canning may reduce residues.
Aldicarb	Residues permeate the produce and probably cannot be removed with washing. Cooking or heat processing may reduce residues.
Azinphos-methyl	Residues remain primarily on the produce surface. Washing, cooking, or heat processing will reduce residues.
Captan	Residues remain primarily on the produce surface. A breakdown product, a suspected carcinogen, may permeate food. Washing, cooking or heat processing will reduce residues.

PESTICIDE	REMOVAL OF RESIDUES
Carbaryl	Residues remain primarily on the produce surface. Washing or peeling will reduce residues.
Chlordane	Residues remain primarily on the produce surface, although there is some evidence that they permeate root crops. No information on removal with water.
Chlorothalonil	Residues remain primarily on the produce surface; however, breakdown products may permeate food. Washing reduces residues.
Chlorpyrifos	Residues remain primarily on the produce surface. No information on removal with water.
DDT	Residues remain primarily on the produce surface, although residues may be absorbed into the peel. Washing, cooking, and commercial processing may reduce residues in some cases.
Diazinon	Residues remain primarily on the produce surface. No information on removal with water.
Dieldrin	Residues remain primarily on the produce surface. There is some evidence that residues permeate crops. No information on removal with water. Peeling or cooking may reduce residues.
Dimethoate	Residues permeate the produce. Washing, peeling, cooking or heat processing may reduce residues.
Diphenylamine	Residues remain primarily in the peel. Plain water washing may not reduce residues. Peeling may help.
Endosulfan	Residues remain primarily on the produce surface. Breakdown products may permeate food. Peeling, cooking or heat processing may reduce residues slightly. No information on removal with water.

PESTICIDE	REMOVAL OF RESIDUES
Ethion	Residues remain primarily on the produce surface. Washing or processing may reduce residues.
Iprodione	Residues remain primarily on the produce surface. No information on removal with water.
Lindane	Some evidence that residues permeate the produce. No information on removal with water.
Malathion	Residues remain primarily on the produce surface, but may be absorbed into the peel. Washing with detergent reduces residues more than plain water. Peeling, cooking, or heat processing will reduce residues.
Methamidophos	Residues permeate the produce and probably cannot be removed with washing.
Methyl Parathion	Residues remain primarily on the produce surface, but there is some evidence that residues can be absorbed. No information on removal with water.
Mevinphos	Residues permeate the produce and probably cannot be removed with washing.
Parathion	Residues remain primarily on the produce surface. Washing, peeling, cooking, or heat processing may reduce residues slightly.
Permethrin	Residues remain primarily on the produce surface. Washing with detergent will reduce residues; plain water may not.
Phosmet	Residues remain primarily on the produce surface. Washing or cooking will reduce residues.
Thiabendazole	Residues are primarily found in the peel. Peeling or washing will reduce residues.

Note: Pesticides in **bold type** are especially hazardous. Many of these are proven or suspected carcinogens (cancer-causing agents).

Adapted from: *Pesticide Alert: A Guide to Pesticides in Fruits and Vegetables* by Lawrie Mott and Karen Snyder. Sierra Club Books, 1987.

BUYING ORGANIC

If you are still concerned about pesticides on your food, especially those that cannot be removed by peeling or washing, you can reduce your chances of pesticide exposure by growing or purchasing organic foods, which are grown and processed without the use of synthetic fertilizers or pesticides. However, because pesticides do not respect boundaries, a neighbour's spray can still drift onto organically grown produce.

Organic foods usually cost more – even twice as much – as foods grown with conventional methods. There are a number of reasons for this:

- Growing crops organically is labour-intensive. Fields may have to be hand-weeded instead of sprayed with poisonous herbicides. Sometimes insects and other pests must also be removed by hand.
- The organic farming industry is not as heavily subsidized or supported in research by government as the mainstream farming industry.
- Land that has been used for conventional farming must go through a transitional period before it can be classified as organic. It takes time for the soil to rid itself of chemical residues and for a natural fertility to build up again through the application of compost, manure and other natural materials. During this time the farmer loses income.

ORGANIC WINES

Wines made from organic grapes don't always bear the word "organic" on their labels. For a free catalogue and information on locating wines from organically grown grapes, call the Organic Grapes Into Wine Alliance at 415-433-0167, or write them at 54 Genoa Place, San Francisco, CA 94133, U.S.A.

- In some regions, demand for organic products continues to exceed supply. This means that producers can afford to charge more. As more farms become organic and there is a greater supply of organic foods, prices likely will stabilize and become more competitive.

How Can You Be Sure It's Organic?

Producers are not allowed to certify their own food products as organic. This must be done by an independent agency. There are at least thirty-nine different organic certification agencies world-wide, including sixteen in Canada. All of them independently set their own standards and may differ in terms of their requirements for certifying products as organic. (Uniform minimal standards for Canada are currently being established.) But there are a few basic standards to which they all comply. Growers must conduct regular soil tests and keep records of their farming practices and the materials they use on their crops. At least once each year an

ORGANIC CROP IMPROVEMENT ASSOCIATION

The Organic Crop Improvement Association is the largest independent certification agency in Canada. It is an international organization and has chapters in every province except Newfoundland. Some of its standards for certification are:

- The land must be free of herbicides (which kill plants) for three years.
- The land must be free of chemical (synthetic) fertilizers for two years.
- Only natural herbicides (plants that are toxic to other plants) may be used to control plant diseases.
- Fields that are vulnerable to drifting chemicals from neighbouring farms will not be certified.

inspector from the certification committee visits the farm to ensure that everything is run according to standard. Soils and crops may be tested for chemical residues. Usually certification means that no pesticides or synthetic fertilizers were applied to the food crop itself and that none have been used on the soil during at least the previous three years.

You can be sure that the food you are buying is organic only if it is accompanied by the logo of a certification agency. A handwritten sign above the produce in the grocery store which says "certified organic" is not enough. It must be clear to you which agency has granted the certification. Look for the name and trademark.

COATINGS ON PRODUCE

Fruits and vegetables typically contain 80 to 90 per cent water by weight. Once they are harvested they quickly begin to lose water and start to shrivel and wilt. They take on an undesirable texture and also lose nutrients. Protective coatings are used to prevent produce from losing water and to maintain their quality. Coatings also prevent the growth of moulds and thus give produce a longer shelf-life. Coatings are also applied because they make food look more attractive. We are more likely to buy shiny apples than dull ones as the shine intensifies their colour.

Coatings not only seal in moisture, they also seal in pesticides. Sometimes coatings specifically contain fungicides to prevent the growth of fungal bacteria. As with pesticides, the Health Protection Branch evaluates residues and sets safety limits for the various coatings.

We are all familiar with picking up a field cucumber or turnip in a store display and feeling the wax coating. But a number of other fruits and vegetables may also be coated. According to the Fresh for Flavour Foundation, a Canadian organization representing the fruit and vegetable industry, these include:

Apples	Melons
Avocados	Nectarines
Bell peppers	Oranges
Cantaloupes	Passion fruit
Eggplant	Peaches
Grapefruit	Pineapples
Lemons	Rutabagas
Limes	Squash

Coatings are also used on some candies and medicinal tablets.

In commercial application, various active ingredients are mixed with emulsifying and wetting agents and water to provide a thin film of wax on the produce. The *Whole Food Bible* describes the various compounds that coatings may contain:

Carnauba wax is obtained from the wax palm of Brazil. It is the hardest of natural waxes and is also used in floor waxes, polishes and lubricants.

Shellac is obtained from the bodies of a female scale insect. Shellac is used as varnish, as a coating on wood and plaster, in electrical insulation and in sealing wax.

Paraffin is derived from petroleum. Because it is flammable and does not dissolve in water, it is widely used to make candles and serves many industrial purposes.

Candelilla wax is obtained from a reed. This natural wax is common in furniture polishes.

Beeswax is made by honey-producing bees.

Sucrose esters of fatty acids are derived from beef tallow.

Some waxes may be removed from produce by washing – and scrubbing. The best way to remove waxes, though, is to peel the produce. This removes valuable nutrients and is difficult to do with some foods, such as green peppers.

BIOREGIONALISM

Eating foods that are grown in your own area or region is known as "bioregionalism." It's a concept that is catching on – for a number of reasons.

- Locally grown produce is more likely to be fresh. Foods transported long distances before they are distributed to stores may have been harvested days, even weeks, before you buy them.
- Chemicals, coatings and various storage methods are often used to keep foods *appearing* fresh. Fewer chemicals are used on produce that does not have to travel long distances before it reaches the consumer. For example, tomatoes to be shipped are picked while hard and green to prevent bruising during transportation. At their destination they may be gassed with an ethylene spray to turn them red.
- Imported fruits and vegetables may be treated with chemicals that are banned in Canada.
- When you buy directly from growers you are often supporting family-run farming operations.

9

ANTIBIOTICS AND HORMONES IN ANIMAL PRODUCTS

ANOTHER CONTROVERSIAL issue surrounding food and its safety is the use of antibiotics and hormones in raising food-producing animals. As is typical of so many other food issues, government, food producers and industry associations continually assure us that our food supply is safe and that the use of hormones and antibiotics in raising animals for food is no cause for concern. Meanwhile a number of consumers and non-governmental groups, and even some livestock producers, remain sceptical and decidedly unassured.

The term "antibacterial" describes any drug that kills bacteria or inhibits their growth. Antibacterials used in raising animals for food include naturally derived antibiotics such as penicillin, streptomycin and tetracycline; synthetically produced substances such as sulpha (sulphonamide) drugs; and a third class of drugs known as nitrofurans. Hormones are chemical compounds that promote growth by causing animals to gain more weight more quickly. Hormones do not treat or prevent disease.

One estimate suggests that one-third of all antibiotics manufactured in Canada each year are used for livestock. In North America, most commercially raised food-producing animals receive drugs of one kind or another during their lifetime. There are basically three reasons why drugs are used in livestock

TABLE 6 Use of Antibiotics and Hormones in Animals

ANIMAL	ANTIBIOTICS ADMINISTERED DIRECTLY	ANTIBIOTICS ADMINISTERED IN FEED	HORMONES
beef cattle	yes	yes	yes
veal calves	yes	yes	yes
dairy cows	yes	yes	yes
poultry (including laying hens)	yes	yes	no
turkey (including laying hens)	yes	yes	no
pigs	yes	yes	no
sheep and lambs	yes	yes	no

production: to treat disease (antibiotics); to prevent disease (antibiotics); to promote growth (antibiotics and hormones).

When high doses of drugs are used to treat disease this is called "therapeutic use" and is typically handled by a veterinarian. When drugs are used at lower dosages to prevent disease and promote growth, the dose is described as "subtherapeutic" and is usually the responsibility of the livestock producer.

ANTIBACTERIALS

Food producers add antibacterials to the animal's feed or drinking water, implant pellets under the animal's skin or inject the drugs directly into the animal's bloodstream. *Canadian Consumer* magazine reports an estimate that 40 per cent of the antibiotics manufactured in North America end up as additives in animal feed. According to Dr. A.V. Rao, Department of Nutritional Sciences, University of Toronto, "Intelligent use of antibacterials in combination with hygienic animal husbandry practices do not pose any health concerns. On the contrary, in some cases, if antibacterials are not used, meat products become unsafe due to the onset of diseases in the animals." One estimate is that, without the use of antibiotics, 15 per cent of animals would become sick during the stressful transfer from ranch to feedlot. Most antibiotics are added to feed by commercial feed manufacturers. These companies must comply with federal government regulations as outlined in the *Feeds Act and Regulations* administered by Agriculture Canada. Farmers regulate themselves in how much feed they give the animals and also in administering the use of antibiotics purchased over the counter. Some provinces require that stores selling these drugs be licensed and keep records of their sales to farmers.

All drugs come with instructions that clearly state the allowable dosage and the required withdrawal period. The withdrawal period is the amount of time, measured in hours, days or weeks, that must pass after the drug is administered before the animal can be slaughtered, in order to allow the drug to be expelled from the animal's body. In some cases, traces of the drug remain in the animal's flesh. Just as for pesticides, Health and Welfare Canada has set allowable maximum residue limits for antibacterial drugs used in meat and poultry production. For many antibiotics, no residues are allowed to remain in the food. Residue limits, along with withdrawal periods, are established by Health and Welfare Canada's Bureau of Veterinary Drugs under the legislation of the Food and Drugs Act. Agriculture Canada is the main body responsible for monitoring residues in livestock.

Use of drugs in farmed fish is uncontrolled. No safe residue limits have been established, and no legal withdrawal periods are enforced.

Opponents of the use of antibiotics and hormones in food-producing animals are concerned with drug residues and the development of drug-resistant bacteria.

Drug Residues

Most antibiotics or hormones administered to an animal are eliminated in the animal's urine or faeces or are broken down in the body into inactive by-products. It is possible, however, for some of the drug to remain in the animal's tissues. Levels of these chemicals are reduced during processing and cooking. However, chemicals may still be present in the edible portions of the flesh and we eat them when we eat their meat.

Residues can cause problems for people who are allergic or in other ways hypersensitive to the drugs. Reactions can range from a mild rash to death. Some of the drugs, such as penicillin (a drug to which many people are allergic), are the same as those administered to humans for treating disease. However, few reactions have been unquestionably linked to these residues in food. In Canada, only one hypersensitivity case has ever been directly linked to antibacterial residues. In 1979 a boy died after eating a hot dog which contained residues of penicillin. But it is possible that many symptoms of reactions to drug residues in food are mistakenly attributed to some other cause.

There is also concern that some of the drugs, such as sulphamethazine, may be carcinogenic. Although no instances of cancer have been directly linked to this compound, it is known to be carcinogenic in rodents. Nitrofuran-based drugs, also known to cause cancer in animals, were only taken off the Canadian market in February 1993. As with almost all other chemicals, little is known about how they might react with the multitude of other chemicals we encounter daily.

Drug-Resistant Bacteria

Antibiotics are used to treat disease both in humans and in livestock animals. They work by weakening or destroying the bacteria that are causing the infection. They also kill many other "non-target" bacteria that are sensitive to the drug.

When an antibiotic is used too often, the bacteria can become resistant. That is, they genetically adapt to the drug and continue to survive despite its presence. Since the antibiotic kills or inhibits all bacteria sensitive to it, whether or not they are the target bacteria, the resistant bacteria can thrive without the competition of other bacteria that would otherwise inhibit their growth. In addition, the resistant bacteria can transfer the genes for resistance to other bacteria of the same or different species. To complicate matters further, bacteria can become resistant to more than one antibiotic.

When bacteria become resistant to a drug in livestock production, farmers must give more of the drug to their animals to keep them healthy or use another antibiotic, until bacteria develop a resistance to that drug as well.

When a species of bacteria develops a resistance to an antibiotic in animals, the bacteria transferred to people will be similarly resistant to that antibiotic. As an example, suppose a strain of bacteria responsible for causing salmonella poisoning becomes resistant to the antibiotic used to treat the condition in livestock animals. When a human contracts the infection (from eating infected meat) the antibiotics used are the same as those used in animals. Once the bacteria are resistant to the antibiotic, it is useless in fighting the infection – both in animals and in humans. Physicians and their patients may find themselves helpless in the face of what once was a "simple" curable infection.

There is disagreement on whether the use of antibiotics in rearing food animals is the main source of resistant bacteria in humans. Many people believe that doctors are too quick to give their patients antibiotics when a health problem arises and feel that it is the over-prescription of antibiotics to human beings, rather than to livestock, which leads to bacterial resistance in humans.

HORMONES

Canadian Living magazine (September 1989) reports that there has been a four-fold increase in sales of hormones used for livestock in the past ten years. Approximately 95 per cent of all beef animals are implanted with hormones to increase growth and feed efficiency.

Hormones are only given to cattle and farmed fish. It is illegal to give animals hormones by way of injection and to implant hormone pellets in areas of the animal that may be consumed by humans.

Growth-promoting hormones are placed under the skin of the ear in beef and veal cattle, and the hormone gradually enters the animal's bloodstream. According to the Canadian *Food and Drug Regulations* the following five hormones may be used for beef and veal: estradiol, progesterone and testosterone (all naturally occurring substances), and zeranol and melengestrol acetate (both synthetically produced).

Livestock producers use growth-promoting hormones to lower their operating costs. Using hormones allows a higher return on feed, speeds up the growth cycle so that more animals can be born, raised, sold and killed in any specified length of time, and increases the selling weight and, therefore, price of each animal. The Beef Information Centre, an industry group sponsored by the Canadian Cattlemen's Association, quotes an estimate that, without the use of hormones, the average price of beef would increase by 15 per cent or more.

In their promotional literature the Beef Information Centre compares hormone levels in beef with those in plant products. They say, "Hormones are found in all foods, whether it be of plant or animal origin. In fact, many plants contain much more estrogen than meat. A pint of beer contains as much estrogen as is found in 100 kg (220 lbs) of beef." But such comparisons are inappropriate. According to Dr. Peter Saschenbrecker, a toxicologist and chief of chemical hazards at Agriculture Canada's Agri-Food Safety Division, estrogen from plants is less stable than that from animal sources and breaks down more easily in the digestive tract.

With the exception of melengestrol acetate, all hormones approved for use in beef cattle in Canada are considered to not pose any hazard by: the World Health Organization/United Nations Food and Agricultural Organization (WHO/FAO) Joint Expert Committee on Feed Additives (JECFA); the U.S. Food and Drug Administration; the Lamming Committee, a European Economic Community (EEC) working group of scientists. The JECFA and the U.S. Food and Drug Administration also consider melengestrol acetate to be free of hazards.

DRUGS FOR ANIMALS – SAFE FOR HUMANS?

Some antibiotics, including penicillin, streptomycin, tetracycline and sulphas, may cause allergic reactions in some people. According to Gerald Guest, Director of the U.S. Food and Drug Administration's Center for Veterinary Medicine, "almost every antimicrobial is potentially dangerous to those individuals sensitive to the substance." Even very low levels pose a risk of hives, fever or shortness of breath to some people, although such cases are rare.

Studies by the U.S. FDA show that sulphamethazine causes cancer in mice and rats. Residues of the drug in animal products could increase slightly the cancer risk for humans. Other sulpha drugs might be cancer-causers, but most of them haven't been tested. Nitrofurazone is another antibiotic for which, according to the U.S. National Toxicology Program, there is "clear evidence" it causes cancer in animals.

Adapted from: Lefferts, Lisa Y. "Cows on Drugs?" *Nutrition Action Healthletter*. April 1990, pp. 8-9.

INSPECTION

Agriculture Canada claims that at federally registered plants "a veterinarian or highly trained inspector examines every animal before slaughter to ensure that it is healthy. . . . After slaughter, our personnel inspect every carcass and all internal organs." More than 2.6 million tonnes of meat are inspected by 1,500 personnel each year. If injection marks are found on an animal its carcass undergoes further testing and analysis.

RESIDUE TESTING

Hormones and antibiotics not approved for use in Canada are not allowed to be present in *any* food animals to be sold in Canada, including imports.

In 1990-91 Agriculture Canada tested 11,593 random samples of beef, veal, mutton, pork, horse, chicken, and turkey for antibiotic residues. All but two samples were free of residues exceeding the maximum allowable limits. (Excess residues of penicillin were found in samples of horse kidney and muscle). In addition to random sampling, Agriculture Canada also specifically samples shipments which they have evidence to suggest may contain residues. In 1990-91 they tested 29,084 total suspect samples of beef, veal, mutton, goat, pork, horse, chicken and duck for antibiotic residues. Of these samples, 770 (2.6 per cent) were found to contain residues in excess of the maximum allowable limits.

Agriculture Canada also tested for excess residues of sulphonamides. No residues of sulphonamides are legally allowed to remain in food. Agriculture Canada randomly sampled pork, veal, beef, mutton, chicken and turkey. Out of 59,013 samples, 248 (0.4 per cent) were found to exceed the maximum residue limits. Suspect sampling was done for pork, veal and beef. Out of

ONLY HORMONE-FREE MEAT FOR EUROPE

On January 1, 1989, the European Economic Community (EEC) banned the use of hormones in all animals to be consumed by humans in that region. This meant that Canadian and U.S. meats and poultry could no longer be exported to any of the twelve countries belonging to the EEC. Dr. Eric Lamming was appointed chairperson of the EEC's Scientific Committee to study and report on the safety of hormones in animal production. In *Meat Probe*, an industry publication, Dr. Lamming writes "The EEC ban on hormones was instituted for political purposes, notwithstanding the scientific evidence. The European hormone ban was clearly established hopefully as a political solution to problems of food surpluses generated by the EEC's Common Agricultural Policy (CAP)." Others disagree with Lamming's analysis and believe that the EEC was taking seriously the concerns of consumers who didn't want to eat hormones in animal products.

8,737 suspect samples, 226 (2.6 per cent) were found to contain residues beyond the maximum allowable limit.

Tests are done for hormone residues of zeranol and melengestrol acetate (MGA), both of which are licensed for use in Canada, and for diethylstilbestrol (DES), trenbolone acetate (TBA) and clenbuterol, none of which are allowed for use in food animals to be sold in Canada. In 1990-91 no violations were found in any of the 2,139 randomly selected samples.

Eggs are also tested for drug residues. Of 3,113 samples tested for antibiotics, all were found to be in compliance. Of 3,075 samples tested for sulphonamides, only one violation was found.

NATURAL MEATS

Some food stores offer consumers "natural" beef. This use of the term "natural" is not the same as that given by Consumer and Corporate Affairs Canada (CCAC). According to CCAC's definition, "natural" food "contains no food additives, added nutrients, flavouring agents, incidental additives or contaminants and has not been further processed." The Beef Information Centre claims that, according to this definition, all fresh beef could legally be labelled "natural." There are currently no government regulations on what constitutes natural beef, no verification of claims that may be made and no on-site inspection. Though the specific claims may differ slightly from brand to brand, the "natural" label as it is currently used typically means that the animals have been pasture-fed (rather than raised in feedlots) and have not received either hormones or antibiotics during their lifetimes. Natural beef may also be aged longer than regular beef.

Many Canadian cattle producers are upset by the use of this label because they think that it implies that regular beef is unhealthy or that it contains residues of antibiotics or hormones. Quoted in *Western Report* (January 2, 1989), Dennis Laycroft of the Alberta Cattle Commission said, "Alarmists have got people worried that colour and flavour [of beef] are enhanced with chemicals. That's absolutely untrue." In *Marketing* (April 4, 1988), Carolyn McDonell of the Beef Information Centre said, "We feel that the thing that's most misleading is when claims are made for substances that don't exist in the product anyway – things like colouring, artificial ingredients and so on. There's no colouring or artificial ingredients in beef." The label of Natural Choice, a brand of "naturally" raised beef offered by Loblaws grocery stores, doesn't make any such claims. What it does say is, "Some consumers remain concerned about possible residues in beef from pesticides, antibiotics and growth stimulants. . . . No evidence exists to indicate that residues are present in conventionally raised beef. However, it is clear that a segment of the market is looking for a naturally raised product."

CERTIFIED ORGANIC MEATS

The term "natural" should not be confused with "certified organic." For beef to be certified organic, producers must not give their animals antibiotics, growth hormones or vaccinations of any kind (with the exception of certain vaccinations that are required by law), and they must provide for their animals feed that is 100 per cent organically grown.

Animals on organic farms are often raised in more comfortable conditions than those raised on conventional farms. In an interview with *The Toronto Star*, Carl Cosack, who raises organic beef at Peace Valley Ranch in Honeywood, Ontario, describes some of these conditions. According to Cosack, "It's very important to eliminate stress among cattle. The government requires that farmers provide no less than 18 square feet per animal in their barns. At Peace Valley, we provide 100 square feet apiece. We bed them with straw, not on slatted boards. We keep them in small groups and don't let the groups mix, so that our cattle are not stressed by having to deal with strangers." The cattle are allowed as much daylight as they like, and swallows are allowed in the barns to keep the fly populations in check. Cosack supposes that these conditions account, at least in part, for the fact that his organic herd is freer of illness than conventionally raised animals. His philosophy is, "As a farmer, you really must ask yourself: 'Am I creating a lifestyle for me – with milking machines and restricted barn movement – or for my cattle?'" Certified organic meats can be found in most large natural food stores.

FISH FARMING

British Columbian salmon farms supply much of the salmon that Canadians eat. In recent years there has been concern about the drugs used in salmon farming. Antibiotics are used because crowded conditions make the fish more susceptible to infections, hormones are used to promote growth, and compounds are used to make the colour of the flesh look more appealing to the consumer. Unlike the regulations for meat production, there are no controls for the use of these drugs in farmed fish. Neither maximum drug residue limits nor withdrawal periods have been legally established.

Conventional fishermen are opposed to fish-farming operations for a number of reasons. Many fear that farmed fish, highly susceptible to disease, might escape and infect wild fish stocks. They are also unhappy that no mandatory labelling exists to allow consumers to distinguish farmed from wild fish. The fishermen's concern, of course, is that consumers may stop buying fish altogether from fear that it contains drug residues. Conventional fishermen also point out that lower grades of farm salmon might turn away consumers of the wild product as well.

A task force of the B.C. branch of the Consumers' Association of Canada studied fish production in British Columbia. In 1989 they issued a report that included the following recommendations (as related in *Canadian Consumer* magazine, 1989, No. 8):

To consumers:

- Insist upon knowing whether you are eating farmed or wild fish and lobby for labelling at the point of sale and on restaurant menus.
- Lobby the Canadian General Standards Board (Ottawa, ON, K1A 1G6) for approved standards for use of antibiotics and other drugs used in fish farming.
- Report any suspected cases of allergic reaction to drug residues in fish to your physician and to: Health and Welfare Canada, Drug Adverse Reaction Program, Health Protection Branch, Ottawa, ON, K1A 0L2.

To government:

- Health and Welfare Canada should establish tolerance limits for drugs and chemicals in fish, and ensure that no fish flesh available on the market exceeds the tolerance limits.
- The federal Department of Fisheries and Oceans and/or the British Columbia Ministry of Agriculture and Fisheries should develop an effective monitoring system to ensure that no drug residues are present in market-ready fish.
- Consumer and Corporate Affairs Canada should change their packaging and labelling legislation and regulations to require the words "farmed fish" or "wild fish" and the harvesting date.

To industry:

- Through the Canadian General Standards Board, salmon farmers should establish farm practices that will reduce the need for the use of antibiotics. One way to do this might be to lower the density of fish in pens.

MILK

Ensuring the safety of the milk we drink is a provincial, not a federal, responsibility. While specific procedures and regulations may vary slightly from province to province, the inspection process is generally the same. Whenever milk is picked up from the farm by a milk transporter, the driver inspects the tank of milk to ensure that it is at the proper temperature and appears clean and odour free. If not satisfied, the transporter can decide to reject the entire tank. The transporter also takes a sample of milk from each tank and sends it to the provincial government for analysis to detect unsafe bacteria levels, added water and the presence of antibiotics. There is a zero-tolerance level for antibiotic residues in milk; none are allowed to remain. In some provinces, such as Quebec, mobile testing laboratories are used to get results within a couple of hours. Tainted milk can thus be dumped before it is shipped to the dairy for testing and processing there.

Farmers are also required to maintain their buildings, equipment and personnel according to set sanitation standards. If

BST: A CANADIAN CONCERN?

Recombinant bovine somatotropin (BST) is a naturally occurring hormone that has caused concern in recent years. BST can be synthetically manufactured and injected into dairy cows to increase milk yields by up to 20 per cent. It is not approved for use in Canada, but has been used here in clinical trials. During these trials, health officials concluded that the milk did not pose any danger to human health and allowed the treated milk to be pooled with regular milk. This meant that consumers in British Columbia, where the trials took place, drank BST-treated milk without their knowledge. When this was publically revealed in 1989, consumers there protested. The Consumers' Association of Canada made a strong statement at that time about the consumer's right to be informed about what is being done to the food supply and why.

Milk from BST-treated milk cannot be distinguished from other milk. Unless it is labelled as treated or untreated, consumers will have no way of choosing between products as there are no tests available that detect differences between milk that contains the naturally occurring hormone and that which contains the synthetic version.

A panel of scientists and physicians brought together by the U.S. National Institute of Health unanimously declared in December 1990 that both the milk and meat of BST-treated animals were safe for human consumption. Still, neither the U.S. Food and Drug Administration nor Health and Welfare Canada has approved the drug for routine use. And given public concerns and fears about the drug's use, it isn't likely to be permitted in the near future.

these standards are not met, or if the milk is found to violate any regulations, the government may choose to take away the dairy farmer's licence to ship milk and may also impose a fine.

When the milk shipment arrives at the processing plant, the processor inspects the milk and can decide to refuse a load from a driver if it seems unfit. Before the milk is pasteurized, the processor tests it for the presence of antibiotic residues. With the exception of milk used for making raw milk cheddar cheese, all milk is pasteurized to kill harmful bacteria. The bacteria required for the manufacture of cultured foods such as cheese, buttermilk, yogurt and sour cream cannot grow in milk that contains residues of antibiotics.

With so many tests and inspections, it would seem that we are virtually guaranteed a safe milk supply. This may not be the case however. In February 1988, *Food In Canada* magazine reported the results of a study at the University of Guelph. David Collins-Thompson, a microbiologist at the university, had tested samples of milk from four provinces using the Charm II test. This test is reputed to be much more sensitive than the tests commonly used to monitor milk, though a limited number of milk processors do use it regularly. Collins-Thompson found that two of the fifty samples contained traces of penicillin and almost 25 per cent had traces of tetracycline and sulphamethazine. All levels were reported to be too low to cause an allergic reaction. Concern remains, however, about the effects of repeated exposure even to such low levels of antibiotics. Critics fear the development of resistant bacteria, which would make antibiotics ineffective in treating humans.

Milk falls under provincial jurisdiction, but Agriculture Canada is responsible for ensuring that other dairy products are free of potentially harmful residues, including antibiotics and sulphonamides. In 1990-91 they tested evaporated milk

WHAT YOU CAN DO ABOUT . . .

DRUG RESIDUES

Although Agriculture Canada rarely encounters violations of the maximum allowable residue limits, some critics question the sensitivity of the tests, the safety of the tolerance levels and the adequacy of sampling sizes. For those who wish to take precautions, the following tips will help you to minimize exposure to hormones and antibiotic residues.

- Drug residues (as well as pesticide residues and heavy metals) tend to concentrate in the fatty tissues of animals. Choose lean meats and low-fat dairy products.
- Reduce your consumption of animal products. Use meat as a supplement to a meal, such as in a vegetable stir-fry, rather than as the main component.
- Cut off all visible fat before cooking meat or poultry.
- Limit your consumption of kidney and liver – residues accumulate in these organs.
- Remember that meat labelled "natural" does not mean that it is "certified organic."
- If you choose organic meat, look for the logo of the certification agency. And, although it isn't clear why, keep in mind that organic beef tends to cook faster than conventional beef.

BACTERIA

Both drug-resistant and non-resistant strains of bacteria are killed by proper cooking. Cooking meat and poultry to the rare stage, a temperature of 60°C, destroys bacteria on the meat surface. However, in the case of ground meat, bacteria may be present throughout. In the case of poultry, salmonella bacteria may survive this temperature, and it is recommended that it be cooked to 77°C, the well-done stage.

(39 samples), milk powders (466 samples), whey powders (57 samples), butter (25 samples) and ice cream (2 samples) for antibiotics. No violations were found in any case; all were free of detectable residues. Sulphonamides were tested for in evaporated milk (17 samples) and milk powders (74 samples). Again, all samples were free of detectable residues.

10

PACKAGING

FOOD ADDITIVES aren't only added directly to food itself, they may also be a part of the packaging. For example, butylated hydroxytoluene (BHT) is sometimes intentionally added to food packaging materials from where it migrates into the food and protects it from spoilage. In much the same way, unintentional additives to packaging may also migrate into food.

In the early 1970s, polychlorinated biphenols (PCBs) were found as contaminants in the cardboard used to package some food products. And let's not forget about the transfer of lead into foods and drinks from lead-soldered cans, lead-based inks used to print bread bags and lead wrappers around wine corks. Also, the growing use of plastics as food packaging materials, particularly for microwave cooking, raises concerns regarding its toxicity. Using plastic wraps, plastic containers and specially-designed plastic packaging may leave unwanted materials in our poultry, peas or pizza (among other foods).

REGULATIONS

According to the *Food and Drug Regulations*, "No person shall sell any food in a package that may yield to its contents any substance that may be injurious to the health of a consumer of a food." In other words, it is illegal to sell food in a package that releases toxins into the food. Simple enough. Or is it?

There seems to be considerable disagreement about the dangers associated with the packaging chemicals that may leach into food. With two exceptions, the federal government has not set acceptable levels or established limits for chemicals that migrate into foods. (The two exceptions are vinyl chloride and acrylonitrile. According to the *Food and Drug Regulations*, neither of these may be present in foods in any detectable amounts as a result of packaging.) The regulation is vague and requires only that judgement calls be made by food manufacturers as to what chemicals, and at what levels, are "injurious to the health of the consumer."

PLASTICS

Plastics are made up of simple molecules called monomers, joined under heat and pressure to form long chains called polymers. The length of the polymer determines the strength of the plastic. Longer chains result in a more durable product.

PLASTICS AND THEIR USES

The Society of the Plastics Industry (SPI) developed a coding system to make it easier to identify the various types of plastics used for containers. Although the system was designed originally to help recyclers sort these materials, consumers, too, can benefit from understanding the numbered symbol.

The Environment and Plastics Institute of Canada explains "there are over 40 basic families of plastics – and each of these can be made into hundreds of variations. Only six are used to produce the plastic containers and packages most commonly found in the home." The container coding system developed by SPI includes these six plastics plus a seventh and is described in Table 8.

Polymers aren't a problem. If they end up in food, they don't usually break down and aren't hazardous if eaten. Monomers are a different story though. Some are very toxic and, if travelling alone rather than in a polymer chain, can migrate from plastic wraps or containers into your food. Not enough is known about how much of these substances make their way into food.

Monomers are only one potential problem with plastics. A host of ingredients are added to plastics to give them various properties. Plasticizers, catalysts, stabilizers, lubricants, release agents, antistatics, antioxidants and colourants can all leach into food.

Plastic wrap

Plastic wraps, or "cling" wraps, are made up of one of three polymer bases: polyvinyl chloride (PVC); polyethylene (PE); and polyvinylidene chloride (PVDC), a copolymer of vinylidene chloride and vinyl chloride. The most common wraps are made from PE and PVC.

With the exception of PE, plasticizers are often added to these bases to give the plastic flexibility. A 1987 study by the British government found that one plasticizer in particular, di-(2-ethyl-hexyl) adipate (DEHA), was making its way into foods considerably more often than others. At that time, DEHA was the most commonly used plasticizer in PVC-based wraps. The study found that DEHA migrated into foods when used in microwave cooking, at room temperature and even at refrigeration temperatures. The amount of migration increased the longer the wrap touched the food and the higher the temperature became. The highest levels of migration were found in microwaved fatty foods, such as meats and bakery products, that had been in direct contact with the plastic wrap. But high levels were also present in fatty meats that had been wrapped in plastic but not microwaved. The lowest levels of DEHA were found in non-fat foods such as fruit and vegetables and in foods that hadn't been

HEAVY METAL BREAD BAGS

Do you ever take an empty plastic bread bag, turn it inside-out to get rid of moisture and crumbs and then, still turned inside-out, store food in it? Don't.

Clifford Weisel and his colleagues at the University of Medicine and Dentistry of New Jersey tested the inks used on the labels of bread bags and found that seventeen out of eighteen samples contained lead. The Consumers Union, the largest American consumers group, examined ninety-four bags and came up with similar results. The lead does not leach through the outside of the bag and into the bread. This is not a concern. The problem arises when the label is turned to the inside and food is stored against it. The researchers estimated that if the food was even slightly acidic, containing vinegar or lemon juice for example, in only ten minutes, ten to twenty-five times as much lead could leach into it as you would normally ingest in one day from all other foods combined. Naturally you aren't likely to store vinegar or lemon juice in a bag. But what about that sandwich with mustard, mayonnaise or relish? All of these condiments have a substantial vinegar content. Tomatoes, too, are acidic.

It's great to reuse plastic bread bags. Just be sure to keep them label-side out.

in direct contact with the wrap, even if it was used to cover the container during cooking. A 1987 study of the use of PVC film in retail stores resulted in similar findings. The highest amounts of migration were found in cheeses, baked goods and sandwiches; somewhat lower in cooked meat and poultry; and lowest in fruits and vegetables. As with the British study, migration increased the longer the wrap was in direct contact with the food.

While the health effects of ingesting DEHA are not known, there is some evidence that it causes cancer in mice, but not in rats. According to the British government's Committee on Toxicity (COT), the risk to humans is probably small. But the COT remains concerned that high doses of DEHA and other plasticizers have caused changes in the livers of test animals. The World Health Organization doesn't feel that there is enough evidence to decide whether it causes human cancer. According to Dr. Robert Ripley of Health and Welfare Canada, "there is no scientific evidence that DEHA is carcinogenic in humans."

The question is, do we want to take the chance? In Britain, other plasticizers are replacing the use of DEHA in PVC plastics. But it isn't yet clear whether these substitutes, acetyltributylcitrate (ATBC) in particular, are any safer when used in the microwave.

The solution seems simple enough. Avoid those plastic wraps that are made of PVC and choose those made of PVDC or PE, which contain fewer plasticizers. But it isn't easy to tell which brand is made of which type of plastic because it isn't necessarily stated on the packaging.

So we went to the manufacturers. Here is what we found out.

TABLE 7 DEHA in Plastic Wrap

PRODUCT	TYPE OF PLASTIC	CONTAINS PLASTICIZERS	CONTAINS DEHA
Glad Cling Wrap	PE	no	no
Handi-Wrap	PE	no	no
Reynolds Plastic Wrap*	PVC	yes	no
Saran Wrap	PVDC	yes	no
Stretch 'n' Seal	PVDC	yes	no

* Once inventory is depleted, this product will no longer be sold in Canada.

Stores often use PVC wraps to package meats, fish, poultry and cheeses. Unlike PE, PVC does not allow oxygen to pass through and thus it prevents meat from turning brown. PVC is the only plastic wrap that completely keeps oxygen out.

WHAT YOU CAN DO ABOUT . . . MIGRATION OF PLASTIC FROM CLING WRAPS

- Don't use them. Store food in the refrigerator in reusable plastic containers (with lids) that are specifically designed for this purpose. In the microwave, use plates to cover food in bowls and use bowls, inverted, to cover food on plates. Alternatively, use heat-resistant glass bowls with lids. In each case, leave a little space to allow steam to escape and be careful when handling – covers and lids can get very hot.
- If you do use them, do not allow the plastic wraps to come in contact with fatty foods, like meats and cheeses, at any time – whether in the microwave or in the refrigerator. Keep these in reusable plastic containers that are designed for storage.
- In microwave cooking it is likely safe to use wraps to cover containers of food but less safe to allow them to touch the food. Leave one corner of the dish uncovered while microwaving or poke a couple of holes in the wrap before removing the dish from the oven. This will allow the steam to escape so that you won't burn yourself and will also prevent the wrap from being sucked down onto the food.
- While it is no guarantee of safety, choose wraps that are *not* made from PVC.
- When shopping for meats, cheeses and other fatty foods avoid those packaged in cling wrap. Go to the deli counter and ask them to cut meat or cheese for you and wrap it in waxed brown paper.
- When buying sandwiches from a delicatessen or take-out counter, make sure that they are freshly made and haven't been sitting wrapped in plastic for any length of time.

Household plastic wraps are regulated by Consumer and Corporate Affairs Canada (CCAC). But, according to Dr. Richard Viau of the Product Safety Branch of CCAC, there are no regulations concerning household plastic wraps. Manufacturers aren't required by law to prove the safety of their products.

Plastic Containers

In the age of "reduce, reuse and recycle" many of us proudly and diligently save and reuse the plastic tubs that previously held margarine, yogurt and ice cream. They seem ideal for food storage and they can go right from the freezer or refrigerator to the microwave – but not necessarily safely.

Yogurt containers were made to hold yogurt. Margarine containers were made to hold margarine. None of these types of containers were designed to store any other food than what comes in them. Although it is generally agreed that they are safe for reuse, they haven't been tested to find out what chemicals (and how much) may migrate from the container to the food when it is microwaved. This is a particular concern when the plastic starts to melt.

Polystyrene containers (commonly known by the brand name Styrofoam) used for take-out foods in restaurants and delicatessens are particularly hazardous. They soften and melt quickly when heated and may release chemicals such as styrene,

MICROWAVE-MELTED MARGARINE TUBS

- To be on the safe side, don't use *any* plastic or polystyrene (Styrofoam) containers in the microwave. Even those specifically designed for this purpose may release chemicals when heated.
- Glass cookware seems to be the safest choice for the microwave. Heat-resistant types, such as Pyrex, are ideal since they can safely be used in both the microwave and conventional oven. Glass-ceramic, such as Corning Ware, is an equally good choice.
- If you use plastic tubs for food storage, make sure that the food is cooled before you pack it away. Heat encourages migration of chemicals.

a known carcinogen and cause of chromosome damage. And yet, according to Dr. Ripley, "There is no evidence to suggest that trace amounts of styrene are hazardous. Indeed a British Ministry of Agriculture, Food and Fisheries 1983 report on styrene concluded that 'there is no likely toxicological hazard to man from present levels of styrene in foods.' Based on available data, the Health Protection Branch also shares that view."

Even plastic containers intended specifically for the microwave oven may not be safe. Manufacturers who use the term "microwave-safe" on their plastic containers or plastic wraps are self-regulating. That is, it is up to them (not the government) to decide what is or is not safe for the microwave. There are no government regulations for making this claim.

There may be reason to worry. In 1989 *Canadian Consumer* published a test story on microwave cookware. Many of the brands of dual-ovenables (containers designed for use in both a microwave and a conventional oven) tested were made from thermoset polyester. The magazine later reported that researchers in England had found benzene present in thermoset polyester material used in the manufacture of plastic cookware.

Heat-Susceptor Packaging

Today, some of us find it very convenient to grab a box of prepared microwaveable food from the refrigerator or freezer, pop it in the microwave and have dinner on our plates three to four minutes later. But along with your pizza or waffles or fries you may also be eating some of the packaging.

Microwave ovens aren't able to brown and crisp foods on their own, but "heat susceptors" built into the packaging in which the food is cooked solve this problem. Heat susceptors are thin grey strips or disks of plastic film which have been metallized with aluminum and then laminated to paperboard with adhesive. They absorb microwave energy in the oven and act

TABLE 8 Container Coding System

SYMBOL	PLASTIC TYPE	TYPICAL PRODUCTS	CHARACTERISTICS
1 PETE	Polyethylene Terephalate (PET)	• bottles for soft drinks and other carbonated beverages • bottles for edible oils, liquor, peanut butter, some household detergents and cleaners, toiletries and other clear bottles	• transparent and glossy • bottles and containers don't have seams but have a nub on the bottom • smooth surface • semi-rigid • will not scratch easily
2 HDPE	High Density Polyethylene (HDPE)	• jugs for milk and distilled and spring water • bottles for juice, laundry bleach, laundry and dish detergents, fabric softeners, motor oils, lubricants and antifreeze • some margarine tubs • crinkly grocery sacks	• natural or coloured with a matte (not shiny) finish • slightly waxy to the touch • ranges from semi-rigid to flexible • will not crack when bent
3 V	Polyvinyl Chloride (PVC)	• bottles for mineral water, salad dressing, vegetable oil, floor polish, mouthwash, liquor • food wraps • blister packs	• translucent, transparent or opaque (coloured, usually a high gloss) • bottles often have seams and may have a faint blue tint • very smooth surface

SYMBOL	PLASTIC TYPE	TYPICAL PRODUCTS	CHARACTERISTICS
	PVC *cont'd*		• will form an opaque white line if bent • usually semi-rigid (though can be a film for food wraps) • scratches easily • holds in aromas while allowing products to breathe
4 LDPE	Low Density Polyethylene (LDPE)	• clear bags, trash bags, grocery sacks, dry cleaning and bread bags, pouch packaging for milk, etc. • may be found in some rigid items such as food storage containers and flexible lids, or as coatings on food cartons, wires and cables	• nearly transparent (garment bags) or opaque (trash bags) • may be coloured • low to high gloss finish (grocery sacks and shopping bags) • flexible • will stretch before tearing when pulled
5 PP	Polypropylene (PP)	• battery cases, appliances, pipe, medical containers, carrying handles, bottle caps, some dairy containers and tubs, shampoo bottles, syrup bottles, snack and food wraps/bags for cheese, breads and cereals, drinking straws and container lids (those that won't crack easily when bent)	• transparent, translucent or opaque • clear or coloured • shiny or low gloss finish • smooth surface • semi-rigid • won't scratch

SYMBOL	PLASTIC TYPE	TYPICAL PRODUCTS	CHARACTERISTICS
6	Polystyrene (PS)	• some yogurt cups and tubs, cookie and muffin trays, clear carry-out containers, some vitamin bottles, most convenience food cutlery • typically used only for wide-mouthed containers (no bottles)	• transparent or opaque • clear or coloured • high gloss • smooth surface • won't scratch easily • brittle to semi-rigid • cracks easily when bent
	Foamed Polystyrene or Expanded Polystyrene (EPS)	• foam drinking cups, meat and produce trays, egg cartons, protective packaging and insulation	• opaque • smooth to grainy finish • comparatively thick-walled • lightweight and "fluffy" • snaps easily when bent
7 OTHER	Other	This refers to products made of plastics other than the six most commonly used plastics. These include products made from multiple plastic resins – often in layers or blends – such as microwaveable serving ware, juice boxes, most snack food bags, and squeezable bottles for condiments and jellies.	

Adapted from *Plastics in Household Packaging*, Environment and Plastics Institute of Canada.

like a small frying pan, actually aiding in the cooking of the food.

Polyethylene terephalate (PET) is the type of plastic used to make these susceptors. When PET and other chemicals were approved for use in these packages by the U.S. Food and Drug Administration, it wasn't expected that they would be used at temperatures above 150°C (300°F). But in microwave ovens and inside packages cooked in conventional ovens, temperatures can be much higher than this – as high as 260°C (500°F). Such intense heat can cause chemicals to migrate from the packaging into the food.

In 1988 the FDA's Center for Food Safety and Applied Nutrition tested a variety of heat-susceptor packages by heating oil in them. Every package tested released chemicals into the oil. Plastic solids were present at levels six to ten times higher than previously found. Up to 95 per cent of PET components, called oligomers, migrated into the oil after six minutes' heating. Trace amounts of benzene, toluene and xylene (all known or suspected carcinogens) were also present, possibly from the breaking down of the adhesives. The FDA subsequently asked that various types of safety data on susceptors be submitted to them. Industry associations joined together and began wide-scale testing.

But members of the food packaging industry felt that the use of corn oil in the first testing may not have given an accurate indication of migration levels as it did not represent actual use. That is, people don't usually heat corn oil in heat-susceptor packaging – they heat the food that comes in it. So the FDA changed its test to one that could determine the amount of heat-susceptor packaging components that makes its way into food during regular use. And they still found migration occurring. Several hundred components could potentially end up in foods – at low levels. Results from the industry tests (fifteen companies submitted forty-two packages for study) have also confirmed that heat-susceptor materials can

RECYCLED PLASTICS

As we faithfully recycle our glass, metal, paper and plastic we like to think that it is really going to be used again. But certain problems arise when recycled materials are used in food packaging.

For glass and metal cans there are few problems with contamination since they are melted down at high temperatures. Paper and plastics are another story. Newspaper may contain dioxins and furans. At least one form of dioxin has been shown to cause cancer in laboratory animals. Plastics used to hold non-food items could have been in contact with any number of hazardous substances. Paper and plastic are recycled at low enough temperatures that some of these contaminants could remain.

But industry is working to find solutions. According to the *FDA Consumer* (November 1991), one possibility includes using recycled plastic as one of the layers in multilayered packaging, where it would not come in actual contact with food.

migrate into the food. Benzene was found in four of the samples and these packages were removed from the market. In no other cases were any other harmful substances, including benzene, found to be present in appreciable amounts.

Health and Welfare Canada has been watching the results from the United States very carefully. This is because all heat-susceptor packaging in Canada either comes from there or is made in Canada under American licence. Dr. Ripley has said, "We haven't seen anything that causes us great concern."

Dual-Ovenables

Some prepared foods come in trays which can be used in either a microwave or a conventional oven. These trays don't contain heat susceptors but they may still release chemicals into food. Tests have found that when dual-ovenables are used in a conventional oven, as much PET ends up in the food as it does from

> A free booklet and fact sheets on plastics and cookware safety is available from the Product Safety Division of Consumer and Corporate Affairs Canada. Contact: Communications Directorate, Place du Portage 1, 50 Victoria St., Hull, PQ K1A 0C9. Tel: 819-997-2983

a heat-susceptor package used in the microwave. There doesn't seem to be this same effect when dual-ovenables are used in the microwave.

Again, it isn't clear that consumption of PET is harmful to human health. But, to be on the safe side, you may want to use dual-ovenable trays in your microwave oven only. Transfer food to an oven-proof baking dish before cooking it in a conventional oven.

MAP, VAP and Sous-Vide Packaging

Food packages do more than make the food contained within look appetizing to consumers. They keep the food fresh and protected as it makes its way from the processor to the distributor to the grocery store and, finally, to the consumer. In recent years, three new packaging methods have been developed that can extend the shelf-life of fresh foods weeks and sometimes months without the use of chemical preservatives.

When foods are exposed to oxygen for too long they begin to spoil. These new packaging methods avoid this problem either by replacing oxygen with other harmless gases or by removing it altogether.

Foods such as entrées containing meat, poultry or fish may be heated before they are sealed in packages. With modified atmosphere packaging (MAP), nitrogen and/or carbon dioxide are used to replace air (oxygen) either in whole or in part. MAP is used for red meats in supermarkets and shelf-stable entrées for home

MAP-ED FRUIT

It is estimated that some 20 per cent of all fruit spoils before it even reaches the consumer. Enzymes cause foods to ripen and eventually deteriorate, moulds may develop and cause spoilage. *Foodservice and Hospitality* (January 1991) reported that a team of researchers at the University of British Columbia have devised a way to diminish this rate of spoilage. Using modified atmosphere technology, the team found that it could slow down the fruit's natural process of decay. Sealing the fruit in flexible film packaging and replacing the oxygen with harmless gases prevents the enzymes and microorganisms from doing their work. When the package is opened the fruit starts to age again.

As with other MAP processing, refrigeration is very important for the process to be successful. When used properly, the method's results are impressive. Raspberries can be kept fresh up to three weeks instead of the usual two days. Fresh sliced pineapple can last two months, and apple slices will stay fresh for up to four weeks. If this application of modified technology takes off, consumers will benefit from a longer fresh fruit season and producers will enjoy larger markets.

Let's not forget, though, that all of these solutions to keeping food fresh involve energy use and packaging – packaging that is eventually thrown out and contributes to the environmental waste problem. A preferable solution is to buy locally grown fresh produce, as needed, and to carry it home in a reusable shopping bag.

microwave ovens. With vacuum packaging (VAP), the air is removed but not replaced. These products are kept refrigerated during shipping and storage until they are ready to be used. Some products, when processed in this way and properly stored, have a shelf-life of up to eighteen months.

Like MAP and VAP, sous-vide is a method of food packaging designed to give prepared foods a longer shelf-life without sacrificing either colour or flavour. Sous-vide, which means "under vacuum," is distinguished from MAP and VAP by the fact that fresh, raw ingredients are placed in plastic pouches and oxygen is removed *prior* to heating.

With all of these techniques, the food may be fully cooked during packaging and require only warming before serving. Alternatively, heating may be minimal during packaging and the food may need to be fully cooked prior to consumption. Non-perishable food items such as freeze-dried coffee and milk powder may also be stored under a modified atmosphere to retain freshness.

Unfortunately, these methods are not without risks. The bacteria responsible for causing botulism (food poisoning) and a few other food-borne pathogens may survive in an oxygen-free or modified environment, and may also be capable of proliferating at relatively low temperatures. But common spoilage organisms, bacteria which would normally warn us that a food is spoiled, cannot grow in the oxygen-free environment of these packages. Without these bacteria to give off a strange odour or flavour we may eat a food contaminated with the botulism bacteria without knowing it, until it is too late. Although no cases have been reported in Canada, an air passenger travelling from France to England was stricken with botulism poisoning after eating an in-flight meal consisting of sous-vide packaged foods.

The same risk of botulism exists with canned foods if the cans are not properly processed or if they are damaged. But cans are sturdier packaging than are plastic pouches and tubs. Plastic containers may rupture more easily, letting air and bacteria into the food and causing it to spoil. With careful handling this can be avoided.

Bacteria that are capable of surviving at relatively low temperatures may infest these foods if they not stored in a cold enough environment. Proper refrigeration is therefore most important, but storage problems can occur, knowingly or unknowingly, at any point in production, distribution and

WHAT YOU CAN DO ABOUT . . . MIGRATION OF CHEMICALS FROM HEAT-SUSCEPTOR PACKAGING

- If you're less than thrilled with the idea of having a side order of plastic, harmless or not, with your microwave-cooked meal, discard the heat-susceptor packaging and cook your meal on a microwave-safe plate instead. The results may not be as brown or crispy but this can easily be fixed by popping the food into the oven or toaster oven a few minutes before eating.
- If you do use the packaging be sure to follow the directions for use exactly. Don't exceed the recommended cooking times. Keep in mind, too, that not all microwaves are calibrated in the same way. "High" on one machine may be much hotter than the same setting on a unit from another manufacturer. If the package becomes charred or very brown, don't eat the product. And never reuse the heat susceptors.
- Another alternative is to avoid buying heat-susceptor packaging altogether. Most of these packages are not recyclable. The types of foods that come packaged this way often include a statement on the label like "brown and crisp." Microwaveable pizza, french fries, breaded fish sticks, waffles and popcorn are likely to come in this kind of packaging.
- Don't buy single-serving microwaveables or other convenience meals. One day spent cooking every week or two could provide you with enough home-cooked frozen meals for weeks. Simply take a stew or casserole out of the freezer in the morning and let it thaw in the fridge for the day. By the time you get home at night you can just pop it in the microwave (in a glass dish) and three to four minutes later have dinner on your plate – with assured safety.

storage. For example, temperatures in grocery store coolers and home refrigerators may not always be low enough to prevent the bacteria from multiplying and becoming a problem.

Unless a food under MAP, VAP or sous-vide is fully cooked, it should not be held in a refrigerator beyond the indicated "use by" date. Remember that these packaging techniques by themselves do not "improve" the product – they only retain its existing quality better and longer than ordinary packaging. Of course, if you have a choice between buying fresh locally grown and packaged produce, choose fresh every time.

Since the quality of the raw ingredients is vitally important, at all stages of food production, workers applying these techniques must be properly trained, and all operations must be carried out under sanitary conditions. Consumers, too, need to learn how to safely handle the foods packaged using these techniques. In this way, health risks will be kept to a minimum and we can all have our cake and eat it too – foods with long shelf-lives that don't contain chemical additives.

11

LABELLING

A GENERATION AGO food shopping was much simpler than it is today. People walked into a store, looked for the product they wanted, compared different brands for price and bought the cheapest one. Today, according to a survey conducted by the Grocery Products Manufacturers of Canada, price is still the number one consideration for consumers. But in the aisles of many grocery stores, more and more people can be found peering at the fine print of product labels. Consumers are beginning to compare nutritional and ingredient information as well as price. In an increasingly health-conscious society, many are paying particular attention to such claims as "no added preservatives," "cholesterol-free," "calorie-reduced" and "high in fibre."

Food labels can serve consumers well in providing us with much of the information we need to make wise choices. But this information is of little use if we can't understand it. Several federal departments have prepared educational materials on labelling and related issues, but, typically, these have only been available to those who seek them out. The government has not done enough to educate the general public. Although they have provided the tool (the label), aside from producing a few pamphlets, they have failed to teach people how to make use of it.

How does one decipher the information on a food label?

What has to appear there and what's missing? What exactly do the descriptive claims mean? And who regulates such information? With answers to these and other questions, you can use the information provided on food labels to compare products and to make wise purchases. Read on.

Whose Responsibility?

The Health Protection Branch of Health and Welfare Canada, under the *Food and Drugs Act and Regulations*, legislates many of the issues related to food labelling. For example, the regulations specify which ingredients must be listed on the label, in what order, and what terms may be used. The *Food and Drug Regulations* also specify in which cases manufacturers are not required to provide ingredient information.

Consumer and Corporate Affairs Canada (CCAC) is responsible for enforcing the labelling, packaging and advertising requirements of the *Consumer Packaging and Labelling Act and Regulations* and the *Food and Drugs Act and Regulations*. CCAC also ensures that information in the marketplace is accurate so that consumers can make informed choices. The Health Protection Branch advises Consumer and Corporate Affairs on labelling and advertising matters that are relevant to public health.

INGREDIENT LISTINGS

In their *Guide for Food Manufacturers and Advertisers*, CCAC says: "Consumers are becoming increasingly concerned about the presence of certain substances in their foods, including allergens, preservatives, caffeine, flavour enhancers, etc. It should be noted that the labelling provisions in the Food and Drugs Act and Regulations *do not require* that *all* ingredients, components and substances present in a food be declared. Therefore, ingredient listings very rarely provide complete information regarding every

A WORD ABOUT CHOLESTEROL

Cholesterol is found *only* in animal products. Cholesterol is not found in vegetable oils, fruits or vegetables, cereals, nuts or grains. It is misleading when foods like potato chips and vegetable oils carry "cholesterol-free" claims; these foods *never* contained cholesterol. These claims also tend to perpetuate the misunderstanding that it is the cholesterol we eat, rather than fats and saturated fats in the diet, that plays a major role in influencing cholesterol levels. And just because a food does not contain cholesterol does not mean that it is fat-free. A cholesterol-free food, for example, may be very high in saturated fats and thus may aggravate a cholesterol problem. Look for nutrition information on the label.

substance which has entered into the production of a food, nor do they provide information regarding substances which are present due to physical or chemical transformations in the food. In other instances, the functional common name of an ingredient may include components such as colouring agents or solvents and dispersing agents which need not be declared because they can collectively be declared as 'colour' or they perform a function in the foods to which they are added."

In other words, there is more in the food than appears on the label.

You Won't Find a List of Ingredients on These Food Labels

According to the Canadian *Food and Drug Regulations*, a list of ingredients must appear on the label of prepackaged products. However, certain products are exempt from this requirement. These are:

- prepackaged products packaged from bulk on the retail premises (e.g., peanut butter in plastic tubs)

- prepackaged individual portions of food that are served by a restaurant or other commercial enterprise with meals or snacks (e.g., ketchup, marmalade or butter)
- prepackaged individual servings of food that are prepared by a commissary and sold by automatic vending machines or mobile canteens
- prepackaged meat, poultry and their by-products that are barbecued, roasted or broiled on the retail premises
- alcoholic beverages for which there are composition standards
- vinegars for which there are composition standards

Retailers are obliged to provide ingredient information for these products on request.

Ingredients That Aren't Listed According to Quantity

Generally, ingredients must be listed in descending order of their proportion or percentage of the prepackaged product. Again, there are exceptions. The following ingredients may be listed in any order at the end of the list of other ingredients:

- spices, seasonings and herbs, except salt
- natural and artificial flavours
- flavour enhancers
- food additives
- vitamins
- vitamin salts or other derivatives of vitamins
- mineral nutrients
- salts of mineral nutrients

According to CCAC, additives and enhancers known to have an effect on health, such as MSG, will be listed by their name and ranked according to proportion.

DON'T OVERLOOK . . . SUGAR

Most ingredients ending with "ose" suggest the presence of sugar. All of the following ingredients, and many others not listed here, are aliases of sugar. Contact the Canadian Diabetes Association or Canadian Dietetic Association for a complete listing.

- barbados molasses
- blackstrap molasses
- brown sugar
- corn syrup
- dextrose
- fructose
- fructose syrup
- glucose
- glucose syrup
- glucose-fructose syrup
- honey
- invert sugar
- lactose
- malt syrups
- maltose
- mannitol
- maple syrup
- raw sugar
- sorbitol
- sorghum molasses
- sucrose
- xylitol

Source: Mindell, Earl. *Unsafe at Any Meal.* New York: Warner Communications, 1987.

Ingredients of Ingredients

If an ingredient is made up of more than one component, all components of that ingredient must be listed on the label. For example, in chocolate chip cookies, the ingredient list will likely include such things as flour, sugar and chocolate chips. Since chocolate chips are made up of a number of ingredients, these ingredients will be listed in parentheses following the mention of the chocolate. Again, there are exceptions. For example, if a product contains cheese but the cheese makes up less than 10 per cent of the total product, the label will list cheese as an ingredient but need not indicate what the ingredients are of the cheese itself. The following is a complete list of ingredients, the components of which do not need to be listed on the label:

- butter
- margarine
- shortening
- lard
- leaf lard
- monoglycerides
- diglycerides
- rice
- starches or modified starches
- breads
- flour except when an ingredient of enriched flour or graham flour
- soy flour
- enriched flour
- graham flour
- whole wheat flour
- baking powder
- milks
- chewing gum base
- sweetening agents
- cocoa
- salt
- vinegars
- alcoholic beverages
- food colours
- flavouring preparations
- artificial flavouring preparations
- spice mixtures
- seasonings or herb mixtures
- cheese, process cheese and cottage cheese when the total amount of such ingredients is less than 10 per cent of the prepackaged product
- jams, marmalades and jellies when the total amount of such ingredients is less than 5 per cent of the prepackaged product
- olives, pickles, relish and horse-radish when the total amount of such ingredients is less than 10 per cent of the prepackaged product
- vegetable, animal or marine oil or fat when the total amount of such ingredients is less than 10 per cent of the prepackaged product
- prepared or preserved meat, fish, poultry meat, meat

by-product or poultry meat by-product when the total amount of such ingredients is less than 10 per cent of a prepackaged product consisting of an unstandardized food

- alimentary (nutritional) paste that does not contain any form of egg or any flour other than wheat flour
- vitamin preparations
- mineral preparations
- food additive preparations
- bacterial culture
- rennet preparations
- hydrolyzed plant protein
- carbonated water
- hydrogenated or modified vegetable, animal or marine oil or fat when the total amount of such ingredients is less than 10 per cent of the prepackaged product
- whey, whey powder, concentrated whey, whey butter and whey butter oil

DON'T OVERLOOK . . . SALT

Salt is a source of sodium. The listing "salt" or "sodium," either alone or in combination with another term, indicates the presence of sodium. Watch for the following terms on the label. All are sources of sodium and should be avoided if you are on a low-sodium diet. This list is by no means exhaustive.

- baking soda
- baking powder
- brine for pickles
- monosodium glutamate (MSG)
- monosodium phosphate
- sodium bicarbonate
- sodium
- sodium stearoyl lactylate
- sodium metabisulphite
- sodium chloride
- sodium citrate
- sodium caseinate
- soy sauce

Source: Mindell, Earl. *Unsafe at Any Meal.* New York: Warner Communications, 1987.

- mould culture
- chlorinated water, fluoridated water
- food flavour enhancers
- gelatin

The composition of these foods is standardized and described in the *Food and Drug Regulations*. The food may contain only the ingredients included in the standard for that food.

By What Name Ingredients Are Listed

Ingredients must be listed by their common name – the name by which they are normally known. The exceptions? The ingredients in Table 9 may be listed under a class name as indicated. According to CCAC, "Class name is the way to ensure all ingredients are given meaningful names that are scientifically accurate and recognized by all." One wonders why class names apply only to these ingredients, however, and not to the thousands of others used in foods. Why are consumers deprived of this information? Take colours for example. Only thirty or so colouring agents are allowed to be used in Canada. As they are among the most allergenic food additives, it seems reasonable that they should be listed by name on food labels, yet according to regulation they need only appear under the class name "colour."

NUTRITIONAL INFORMATION

Although not related directly to food additives, nutritional information is another component of the information found on a food label. As it is an area unfamiliar to many consumers, it will be useful to discuss nutritional information and labelling here. (Just a reminder: vitamins and minerals are not considered food additives in the *Food and Drug Regulations*.)

TABLE 9 Class Names of Ingredients

INGREDIENT	CLASS NAME
one or more vegetable fats or oils, except coconut oil, palm oil, palm kernel oil or cocoa butter	vegetable oil or vegetable fat
one or more marine fats or oil	marine oil (i.e. oil from marine animals – whales, seals, etc.)
one or more permitted food colours	colour
one or more natural flavours	flavour
one or more artificial flavours	artificial flavour, imitation flavour or simulated flavour
one or more spices, seasonings or herbs except salt	spices, seasonings or herbs
any combination of all types of milk: whole, skimmed or partly skimmed, cream, butter and butter oil	milk solids or dairy ingredients
any combination of disodium phosphate, monosodium phosphate, sodium hexametaphosphate, sodium tripolyphosphate, tetrasodium pyrophosphate and sodium acid pyrophosphate	sodium phosphate or sodium phosphates
one or more species of bacteria	bacterial culture
one or more species of mould	mold culture or mould culture
preparation containing rennin	rennet
milk coagulating enzymes	microbial enzyme

In Canada it has only been since November 1988 that manufacturers have been legally permitted to label their products with nutrition information. Although ingredient listings have always been mandatory (with some exceptions), nutrition labelling is entirely voluntary. Although companies are in no way obliged to provide nutrition information, the need to remain competitive may force them to do so. In the case of breakfast cereals, for example, it already has. Virtually all cereal boxes were printed with nutritional information almost immediately after the legislation was passed.

The revisions introduced with the 1988 legislation were of two types: one, the way nutritional information may be presented and, two, the claims that are allowed to be made on a food label or in an advertisement.

Recommended Daily Intake

Changes were made as to the way *amounts* of nutrients can be expressed. If a nutrient claim is made, a declaration of the amount of the nutrient claimed per serving of stated size is mandatory. All nutrition information has to be declared about the product itself and not the way it is served. For example, nutritional claims on a cereal box must pertain only to the cereal itself and not to the nutritive value of the cereal served with milk. Values for energy are given in both calories and kilojoules, and protein, fat, carbohydrates and fibre in grams. Sodium, potassium and cholesterol are given in milligrams. Vitamins and mineral nutrients must be expressed as a percentage of the Recommended Daily Intake (RDI) as set by the federal government. For example, if a label reads "Vitamin C: 30%" it means that one serving of that food provides 30 per cent of the vitamin C that an average person needs each day. If the vitamin or mineral nutrient represents less than 5 per cent of the RDI, it may be listed provided no claims related to it are made.

Of course, the RDI is meant only as a guideline to suggest what people need, on average, in terms of nutrients. It is impossible to take into account all the ways in which people vary (age, sex, lifestyle, metabolic and physiological differences) and thus have different nutritional needs. According to Pat Steele of Consumer and Corporate Affairs Canada, "[The RDI] isn't intended to address the individual's nutritional needs. It gives consumers an idea of the nutritional contribution that food will make to their daily diet and serves as a uniform standard to help consumers more readily compare the nutrient value of foods."

Serving Sizes

Changes were also made to the ways that serving sizes are declared on a label. Previously, nutrients were expressed per 100 grams or per 100 millilitres of product. This was helpful in some ways but not so in others. Since nutrients were expressed in terms of the same quantity on all products, it was very easy to compare the nutrient values of different foods. But this comparison was potentially misleading. For example, there may be as much of a particular nutrient in 100 grams of an enriched cereal as there is in 100 grams of peanut butter. But it is more likely that you would consume 100 grams of a granola-type cereal – perhaps 1 cup – as one serving than that you would consume 100 grams of peanut butter – about six tablespoons – at one sitting.

Under the revised regulations, nutrient information must be based on stated serving size. The federal government provides guidelines (not regulations) on what serving size or range of sizes is considered appropriate for various foods. This allows the consumer to make more accurate comparisons between one food type and another because it takes into account the actual amount of that food you are likely to consume as one serving. Or at least it does in theory. There are, unfortunately, a few problems with this system that may prevent comparisons from being as accurate as one would like them to be.

Serving sizes are expressed in metric units first (e.g., grams, millilitres) followed by an equivalent household measure or common unit in parentheses (e.g., cups, two pieces). In order to make accurate comparisons between foods it is important to understand that the nutrition information is based on mass (or weight) or, in the case of liquids, volume. Foods of the same type often, though not always, have the same mass for a single serving size. For example, a single serving of a breakfast cereal, whether puffed rice or fruit-and-nut granola, is set at 30 grams. This is a measure of mass. Although the mass may be the same for foods of the same type, the foods may vary considerably in terms of their volume. You would need much more of a cereal like puffed rice to make up 30 grams than you would a more dense cereal like granola.

The question is whether serving sizes are realistic in terms of the amount you would actually eat at one sitting. In the case of peanut butter, 35 grams (two tablespoons) does seem a reasonable serving size. In the case of granola, 30 grams (¼ cup) does not seem realistic. Neither does twelve potato chips or six pretzels. Most people would eat more than this. In no case does it appear that the serving size *overestimates* what an average person is likely to eat. Underestimates, however, are likely. When a serving size is underestimated the amount of sugar, fat or calories in one serving is also underestimated and may seem nominal. But if you realistically consume two, three or even four times the suggested serving size, the amounts of the less desirable nutrients begin to have more significance. This information may be hidden if you don't pay attention to the serving size as specified and estimate how much you are actually likely to eat.

Table 10 gives some examples of serving sizes as suggested by Consumer and Corporate Affairs Canada. CCAC says: "It should be emphasized that these serving sizes are intended to be suggestions only. Industry and individual manufacturers will have the flexibility to determine the serving size for a given product provided that it is reasonable, and is used in a fair and consistent manner. A reasonable serving size is considered to be an amount of food which would reasonably be consumed at one sitting by an adult."

TABLE 10 Suggested Serving Sizes

FOOD	SERVING SIZE
Pasta, cooked (macaroni, noodles, spaghetti)	125–250 mL (⅔–1 cup)
Rice, cooked	125–175 mL (½–¾ cup)
Dressings for salads	15 mL (1 tbsp.)
Fruit juices, fresh, frozen, canned	175 mL (¾ cup)
Nuts	60–125 mL (⅓–½ cup)
Cheese, cottage	125 mL (½ cup)
Ice cream	125 mL (½ cup)
Soup	175–250 mL (¾–1 cup)
Vegetables, raw	75–250 mL (⅓–1 cup) 1 carrot, 1 stalk celery, ½ cup shredded cabbage, 1 cup lettuce, ⅓ cup green peppers
Snacks	15–40 g 12 potato chips 2 cups popcorn 6 pretzels

Nutritional Claims

In addition to providing a basic nutritional breakdown of their products, many companies are also using descriptive claims to appeal to consumer concern with calories, fat, sodium and sugar. If a claim is made about a particular nutrient, the company must state the amount of the nutrient for which the claim is made.

However, the company doesn't have to reveal anything about the amount of other nutrients in the food. When this ruling was made, many people objected to this fact, preferring a mandatory list of core nutrients whenever any nutrition claim is made. A core listing would prevent companies from highlighting only the positive aspects of the product while failing to reveal any negative features. For example, a product advertised as being low in sugar may be teeming with both fat and sodium. Yet, according to the regulations the manufacturer is obliged only to reveal the amount of sugar in the product.

With the number of claims that abound these days, it can be a challenge to decide which product is the best. If you are counting calories, is it better to choose a product that claims to be "calorie-reduced" or one that is "low calorie"? Table 11 should help you interpret and understand many of the descriptive claims that are made regarding calories, protein, fat, sugar, fibre, salt and vitamins.

OTHER DESCRIPTIVE CLAIMS

In addition to those claims described in the table, many other declarations adorn the labels of the myriad of food products on grocery store shelves. CCAC provides food manufacturers with guidelines on the appropriate use of these labelling terms. Some of these guidelines, as described by CCAC, are outlined below. We have included terms that are frequently encountered and often misunderstood.

Contains no preservatives or *Preservative-free* may be used when the product contains none of the known preservatives. If an ingredient known as a preservative is present, even if for purposes other than preservation, this claim may not be used. The words "Contains no" and " __ -free" can also be made in regard

TABLE 11 What Nutritional Claims Really Mean

THE CLAIM . . .	MEANS THAT THE FOOD . . .
CALORIES	
calorie-reduced	contains at least 50% fewer calories than the same food when not calorie-reduced; for special dietary use foods
low calorie low in energy light in calories lite in calories light in energy lite in energy	is calorie-reduced and contains 15 calories or less per serving; usually contains fewer calories than a calorie-reduced food but more than a calorie-free food
__ % less calories than . . . lower in calories than . . . less calories than . . .	compared to the reference food, contains at least 25% fewer calories and at least 30 fewer calories per serving
calorie-free	contains no more than 1 calorie per 100 g of food
source of energy	contains at least 100 calories per serving as indicated on the label
light dinner lite dinner light meal lite meal	contains no more than 300 calories; contains at least one average-sized serving from either the meat and alternates or milk and dairy products food groups and either the fruit and vegetable or cereal food groups
PROTEIN	
excellent source of protein very high in protein	has a greater quantity and/or better quality of proteins than if the label indicates "contains," "source of," "good source of" or "high in protein"

THE CLAIM ...	MEANS THAT THE FOOD ...
FAT*	
low-fat low in fat light in fat lite in fat	contains no more than 3 g of fat per serving and no more than 15 g of fat per 100 g of dry matter
__ % less fat than ... lower in fat than ... reduced in fat	compared to the reference food, has at least 25% less fat, at least 1.5 g less fat per serving and no increase in calories
fat-free contains no fat	contains no more than 0.1 g of fat per 100 g
low in saturated fatty acids low in saturates low in saturated fats	contains no more than 2 g of saturated fatty acids per serving and derives no more than 15% of calories from these
free of saturated fatty acids free of saturates free of saturated fats	contains less than 0.1 g saturated fatty acids per 100 g food
__ % less saturated fat than ... lower in saturates than ... reduced in saturates	compared to the reference food, has at least 25% fewer saturates and at least 1 g less fat per serving
low-cholesterol low in cholesterol light in cholesterol lite in cholesterol	contains no more than 20 mg of cholesterol per serving and per 100 g, no more than 2 g of saturated fatty acids per serving and derives no more than 15% of calories from these
cholesterol-free free of cholesterol no cholesterol	contains no more than 3 mg of cholesterol per 100 g, no more than 2 g of saturated fatty acids per serving and derives no more than 15% of calories from these

* Fat claims are not permitted for cheese products.

THE CLAIM . . .	MEANS THAT THE FOOD . . .
__ % less cholesterol than . . . lower in cholesterol than . . . reduced in cholesterol	compared to the reference food, has at least 25% less cholesterol and saturated fat, at least 30 mg less cholesterol per serving and at least 1 g less saturated fat per serving
CARBOHYDRATES	
carbohydrate-reduced	would have at least 25% of the calories from its carbohydrate content if it were not reduced and contains at least 50% fewer carbohydrates and no more calories than if it were not carbohydrate-reduced; for special dietary use foods
low-sugar low in sugar light in sugar lite in sugar	contains no more than 2 g of sugars per serving and no more than 10% sugars on a dry basis
__ % less sugar than . . . reduced in sugar lower in sugar than . . . lightly sweetened	compared to the reference food, has at least 25% fewer sugars, at least 5 g per serving fewer sugars and no increase in calories
sugar-free sugarless no sugar sweet without sugar	is a carbohydrate-reduced food that, when ready to serve, contains no more than 0.25 g of sugar per 100 g and no more than 1 calorie per 100 g or 100 mL (except chewing gum); usually contains the least amount of sugar and often the fewest calories; for special dietary use foods
no sugar added unsweetened	has no added sucrose or other sugars (e.g. honey, molasses, fruit juice, fructose, glucose) although it may contain naturally present sugar

THE CLAIM . . .	MEANS THAT THE FOOD . . .
no added sugar, sweetened with . . . sweetened with . . .	contains no added sucrose but may contain other sweeteners such as honey, molasses, fruit juice, fructose, glucose, etc.
FIBRE	
contains a moderate amount of (naming the fibre source, e.g., oat bran) source of dietary fibre made with (naming the fibre source, e.g., oat bran)	has at least 2 g of dietary fibre per serving
high source of dietary fibre high in dietary fibre	has at least 4 g of dietary fibre per serving
very high source of dietary fibre very high in dietary fibre fibre rich	has at least 6 g of dietary fibre per serving
SALT AND SODIUM	
low-sodium low salt low in sodium low in salt light in sodium lite in sodium light in salt lite in salt	contains 50% less sodium than the regular products and not more than 40 mg of sodium per 100 g and no salt has been added; for special dietary use foods Exceptions: cheddar cheese may contain up to 50 mg of sodium per 100 g; meat, poultry and fish may contain up to 80 mg of sodium per 100 g
__ % less sodium/salt than . . . lightly salted	compared to the reference food, has at least 25% less sodium/salt and at least 100 mg less sodium per serving
sodium free salt free	contains no more than 5 mg of sodium per 100 g; usually contains the smallest amount of salt or sodium

THE CLAIM . . .	MEANS THAT THE FOOD . . .
no added salt, unsalted	has not had any salt added and no ingredient or component contributes a significant amount of sodium to the food
VITAMINS AND MINERALS*	
source of . . . contains . . .	contains at least 5% of the recommended daily intake
good source of . . . high in . . .	contains at least 15% of the recommended daily intake (30% in the case of vitamin C)
excellent source of . . . very high in . . . rich in . . .	contains at least 25% of the recommended daily intake (50% in the case of vitamin C)

* Claims may be made only for vitamins or mineral nutrients for which recommended daily intakes have been established.

Source: *Guide for Food Manufacturers and Advertisers*. Consumer and Corporate Affairs Canada, 1992.

to food colours, flavour enhancers, flavouring agents, and other food additives.

Contains no added preservatives means that no preservatives have been intentionally added, but that some substances typically used as preservatives may be naturally present. For example, ascorbic acid (vitamin C) is used as a preservative. Since oranges naturally contain ascorbic acid, a drink made from them cannot carry the claim "Contains no preservatives." If no preservatives have been added directly to the drink or to any of its ingredients, however, it can bear the statement "No added preservatives." (For labelling and advertising purposes, salt, sugar, vinegar and lemon juice are not considered as preservatives.) Again, this claim also applies to food colours, flavour enhancers, flavouring agents, and other food additives.

If a claim is made that a food does not contain any added substance or ingredient, the label should state what amount of that substance or ingredient is *naturally* present in the food when the amount present is significant.

Fresh means that the food has not been frozen or preserved by any method and is being offered for sale at the earliest possible time. Refrigerated foods such as meats, fruits and vegetables are usually considered fresh.

Light or *Lite* has a different meaning according to what part of the food it describes as being "light," and this must be clearly stated somewhere on the label. Claims such as "light texture," "light taste" or "light tasting" do not mean that the food contains less of something such as salt or fat. The claim refers only to the texture or the flavour.

Natural should only be applied to foods that have not been significantly altered physically, chemically or biologically by processing. A natural food or ingredient of a food is one that does not have anything added, such as vitamins or mineral nutrients, artificial flavouring agents or food additives. A natural food or ingredient of a food is also one which does not have anything removed or significantly changed.

Some food additives, vitamins and mineral nutrients may be derived from natural sources and, in some cases, be considered natural ingredients. When this is the case, the claim should be that the food contains "natural ingredients." Remember that simply because an ingredient comes from a naturally-occurring source does not mean that it is safe.

Pure; 100% Pure; 100%; All should only be applied to single-ingredient foods that are uncontaminated, unadulterated and contain only substances or ingredients that are understood to be part of the food so described. For example, a product described as "100% pure corn oil" is expected to contain only corn oil. It should not contain any preservatives, antifoaming agents or colour even though regulations allow these substances to be added to such an oil.

When the above terms are applied specifically to an ingredient name, rather than to the whole product, the relevant ingredient should meet the same guidelines described above. That is, the ingredient should not have anything added to it. When the term "pure," "100% pure" or similar wording applies to an ingredient, it must be clear that the term applies only to the ingredient and not to the entire food product. For example, "pure strawberry jam" means that no fruit other than strawberries is in the jam, but other ingredients may be found besides strawberries, such as sugar and pectin.

No tropical oils means that the product does not contain coconut oil, palm oil, palm kernel oil or cocoa butter. All of these are high in saturated fat. However, a product may still be high in saturated fat if it contains, for example, hydrogenated vegetable oil. Look for nutrition information on the label.

12

IRRADIATION

THERE WAS A time when "nuked food" and "zapped meals" were disparaging terms for food cooked in microwave ovens. Now that most homes have them, fears about microwaves in the kitchen have pretty well subsided. But don't throw away those trendy expressions yet. You may need them if irradiation technology takes hold in the food processing industry.

Irradiation involves exposing materials to gamma rays. Objects are placed in containers, put on a conveyer belt and carried into a chamber constructed of thick concrete walls. The source of the radiation, cobalt-60 or cesium-137, sits on a rack in the chamber. Since the source constantly emits irradiation, it is immersed in a pool of water when not in use. Cobalt-60 is made from mined cobalt metal. Cesium-137 is a waste product of the nuclear industry. Gamma radiation is used in the treatment of cancer patients, to sterilize medical equipment and to extend the shelf-life of some foods. Of all these applications, food irradiation has stirred up the most controversy.

Most of the radiation passes right through the food, but a small amount is absorbed. The radiation dose determines how much radiation is absorbed and, therefore, what the effect on the food will be. Low doses of irradiation can be used to increase the shelf-life of some fruits and vegetables by changing the biochemistry of their cells and slowing ripening and decay. Higher levels kill insects (or sterilize them to prevent reproduction) and destroy microorganisms that cause food to spoil. Food irradiation is meant to be used in combination with other methods of food preservation (such as refrigeration) rather than to replace them.

A "rad" is a measure of radiation that stands for "radiation absorbed dose." The amount of radiation in an ordinary chest X-ray is less than one rad. To kill insects or to sterilize them so they don't reproduce requires 10,000 to 50,000 rads. Fresh fruits and vegetables are exposed to about 100,000 rad; pork to about 300,000 rad; and spices are set at about 1,000,000 rad. Note that 100 rad equals 1 Gray (Gy), and 100,000 rad equals 1 kiloGray (kGy). Health and Welfare Canada has accepted the international recommendation of 10 kiloGray (1,000,000 rad) general clearance. This means that tests to prove safety are required only when the dose of radiation to be applied to a food exceeds 10 kiloGray.

HEALTH AND WELFARE CANADA'S RULES

In June 1989, Health and Welfare Canada changed its regulations to consider food irradiation as a process rather than as a food additive. This change means that toxicity testing is no longer mandatory. But Health and Welfare continues to require submission of a variety of data when an application is made to irradiate a food. This includes studies that show the nutritional value of the irradiated food, and other types of studies that show the food hasn't been significantly altered (physically, chemically or microbiologically). If existing toxicity studies are out of date or inappropriate, Health and Welfare may require that new toxicity testing be completed. But if a previous toxicity test is considered valid, the application may be approved without new research.

FOODS THAT MIGHT BE IRRADIATED

Health and Welfare Canada has approved irradiation for potatoes, onions, wheat, wheat flour and spices. The technology was approved for use on spices in 1984 and for the other foods in the mid-1960s. Poultry and cod and haddock fillets have been cleared for "test marketing" since 1973. But so far the only irradiated foods sold in Canada are spices used in the preparation of other foods. The Canadian Irradiation Centre, part of Nordion International Inc. and located in Laval, Quebec, treats 100 to 200 tonnes of spices each year. This represents less than 5 per cent of total spice consumption in Canada.

TABLE 12 Permitted Radiation*

FOOD	PERMITTED SOURCE OF RADIATION	PURPOSE OF TREATMENT	PERMITTED ABSORBED DOSE
Potatoes	Cobalt-60	To inhibit sprouting during storage	15,000 rad (0.15 kGy)
Onions	Cobalt-60	To inhibit sprouting during storage	15,000 rad (0.15 kGy)
Wheat, Flour, Whole Wheat Flour	Cobalt-60	To control insect infestation in stored food	75,000 rad (0.75 kGy)
Whole or ground spices and dehydrated seasoning preparations	Cobalt-60, Cesium-137, or electrons from machine sources	To reduce the presence and growth of bacteria	1,000,000 rad (10.00 kGy)

* Based on information provided in the Canadian *Food and Drug Regulations*

Under the Canadian *Food and Drug Regulations*, packages of irradiated foods must carry the international food irradiation symbol (called the radura symbol) and must also include a written statement such as "treated with irradiation," "treated by irradiation" or "irradiated."

Unpackaged foods must be accompanied by a sign next to the food with the same information. Ingredients in a food only need to be labelled as irradiated if they make up 10 per cent or more of the total product. A food product could include 9 per cent irradiated onions, 9 per cent irradiated potatoes, 5 per cent irradiated flour and 2 per cent of irradiated spices. Although 25 per cent of the product's ingredients had been irradiated, and theoretically 100 per cent could have been, it would not need to be labelled as such.

IRRADIATION IN CANADA

Michelle Marcotte, a market development specialist for Nordion International Inc., has said that a key reason why irradiation is not more widely practised in Canada is that "we have less of a food processing problem than in other countries, and have less infestation by insects. We have good potato and onion storage facilities, and food storage for grain, with our cold winters." Margaret Barber, of Probe International, disagrees. She has said, "Consumer outrage in Canada stopped the test-marketing a few years ago of irradiated potatoes." Indeed, a 1988 poll by Environics Research Group Ltd. found that 75 per cent of Canadians said no to food irradiation.

IS IT SAFE?

Food irradiation has been a hotly debated subject for years, with scientists and physicians on both sides of the safety issue.

According to Health and Welfare Canada, irradiation has been proven to be safe through more than thirty years of toxicological and genetic testing. Studies by the International Atomic Energy Agency (IAEA) concluded that 10 kiloGrays (1,000,000 rads) of irradiation was safe for any food and use at this level could be approved without further analysis. The World Health Organization, United Nations Food and Agricultural Organization, Science Council of Canada, Agriculture Canada, and Consumer and Corporate Affairs Canada have all endorsed food irradiation. About fifty countries permit the irradiation of various foods. To varying degrees, approximately twenty of these countries are currently doing so. In December 1991, *Report on Business Magazine* related that, in Canada, "only about half a million metric tonnes of food products and ingredients are irradiated, which is a tiny fraction of the volume treated with chemicals and pesticides or by other methods such as vapour heat."

Not everyone supports it, however. Opponents of food irradiation include the British Medical Association, the Canadian Medical Association, and the Consumer Health Organization of Canada. Some critics say that the majority of the studies suggesting that irradiation is safe were published by government departments or the nuclear industry and haven't been properly reviewed by the scientific community. According to Kathy Cooper of the Canadian Environmental Law Association, food irradiation was given regulatory approval based on "inappropriate" data. "Two-thirds of this information was either completely unpublished or published by government departments, atomic energy agencies or food irradiation projects. It hasn't gone through rigorous peer review in the internationally respected scientific literature – which is the test you apply to scientific information to see if it's worthwhile."

Since Canada's regulatory actions are often influenced by those of the United States, it is worthwhile to look at the process

NO IRRADIATION HERE

- Maine, New York and New Jersey all prohibit the sale of irradiated food. A number of other states are considering making the same move. These bans do not include irradiated spices that are used in processed foods since the practice is entrenched and would be difficult to regulate.
- In the U.S., more than 1,000 companies have said that they will not use or sell irradiated products. These include: H.J. Heinz, Perdue, A & P, Quaker Oats, Marks and Spencer, Ralston Purina, McDonald's and Campbell's.
- Abu Dhabi, Australia, Austria, Botswana, Denmark, the Dominican Republic, Ethiopia, Kenya, New Zealand, Rumania, Sweden, Switzerland, Tanzania, the United Kingdom and West Germany have all either banned or severely limited food irradiation.

of approving irradiation there. When the U.S. Food and Drug Administration (FDA) made its decision about the safety of food irradiation, there were 441 toxicity studies available. In her booklet, *The Facts About Food Irradiation*, Irene Kock comments on these studies: "Only 69 were found to meet the criteria chosen for the evaluation. Of these, 37 suggested safety and 32 suggested adverse effects. The FDA's Final Report of the Task Group of Toxicological Data on Irradiated Foods accepted only five studies on which to base the proof for the safety of irradiation. In fact, the broad clearance granted by the FDA in 1986 was not actually based on these five studies, but on a theoretical estimate of the number and amount of radiolytic products likely to be created by irradiation." According to Dr. Bev Huston, chief of Chemical Evaluation for the food directorate of Health and Welfare Canada's Health Protection Branch, the fact that so many studies were rejected by the FDA is no need for concern. "That's not unusual when you deal with a large body of data, much of which is now quite old. The studies were up to thirty years old, remember. That's a long time in scientific terms."

The five studies that the FDA did accept were reviewed at the Department of Preventive Medicine and Community Health of the New Jersey Medical School. Researchers there found two of them to be statistically invalid. One of the two studies suggested that older animals suffered adverse effects after having been fed irradiated food. In a third study, animals fed irradiated food experienced weight loss and miscarriages. Irradiation destroys vitamin E, and a deficiency in this vitamin was the likely explanation for these effects. The two remaining studies were judged to be valid. However, these studies considered the effects of radiation at levels that are below those currently approved for use on some foods. They do not prove the safety of these irradiation levels.

In the mid-1970s the National Institute of Nutrition in India used a group of fifteen malnourished children to study the effects of irradiated food on humans. The experiment was conducted based on the concern that during food shortages, freshly irradiated grains would likely be consumed. Five of the children were fed freshly irradiated wheat, five were fed non-irradiated wheat and the remaining five were fed irradiated wheat that had been stored for at least three months. All the children, except for the group that consumed the non-irradiated wheat, showed changes in their white blood cells. The children who ate stored irradiated wheat developed some polyploid white blood cells. Those who ate freshly irradiated wheat developed larger numbers of abnormal and polyploid white blood cells. As Irene Kock says, "Normal human cells contain 46 chromosomes and are the location of genetic information. Polyploid cells contain one or more extra complete sets of chromosomes. A human polyploid cell could contain 2 sets (92 chromosomes), 3 sets (138 chromosomes), etc. Polyploidy is relatively rare in human cells. Although it is unclear exactly how polyploidy affects humans, cancer researchers have noticed some common traits between polyploidy, and leukaemia, a form of cancer in which there is excess production of white blood cells." It should be noted that the cell abnormalities increased as long as the subjects continued to eat the irradiated wheat, but decreased when the diet was discontinued.

This study has been hotly debated. Critics say that too few subjects were studied, they were a very specific group (that is, malnourished children) and there is some question about the statistical methods used. Health and Welfare Canada similarly

IRRADIATION IN THAILAND

Thanks to a grant from the Canadian International Development Agency (CIDA) of $4.7 million, Thailand opened a new irradiation facility in August 1989. The plant was built by Atomic Energy of Canada Ltd. (AECL). The Nuclear Awareness Project in Oshawa, Ontario, reports that, according to an agreement signed in 1987, the Canadian government is to "help Thailand promote and market their irradiated foods in Canadian stores." These foods include papayas, mangoes and frozen shrimp. There is currently no clearance for the sale of these irradiated items in Canada. Promoting the sale of Canadian nuclear technologies seems to have been CIDA's priority.

Organized by Probe International, an advocacy group which has opposed irradiation for many years, a number of groups warned the Thai government that they will boycott their irradiated goods. This led consumer groups in Thailand to become concerned that the irradiated foods would be dumped on their home markets. The Thai government stated that the irradiated foods would not be sold in Thailand. It is curious that countries that are irradiating food seem more interested in exporting it to foreign markets than they are in serving it at home.

As for CIDA, their commitment to the project includes construction of the facility, tests to see how well the technology works and market trials of the products in Canada.

dismissed the study, noting that the polyploidy may have been linked to the children's malnourishment prior to the study. (The FDA came to the same conclusion.) If it is the case that the malnourished are particularly vulnerable to the dangers of an irradiated diet, and if, at the same time, irradiation is to be used to improve the distribution of food to less developed countries, millions of malnourished people could be placed at risk.

Another study, conducted by the Shanghai Institute of Radiation Medicine and the Shanghai Institute of Nuclear Research in 1987, involved seventy healthy men and women. Results suggested that subjects who ate irradiated foods that had been stored for some time had slightly more chromosomal abnormalities than did a group who ate non-irradiated food. Although neither of these studies provides clear evidence, they do raise some concerns.

NORDION INTERNATIONAL INC.

Canada has been a leader in the development and promotion of food irradiation. Atomic Energy of Canada Ltd. (AECL) has 80 per cent of the world market for cobalt-60. Previously owned by AECL and now a private corporation based in Kanata, Ontario, Nordion International Inc. produces and sells the material. In 1990 its profits were $16.5 million.

WHY IS IT USED?

Pesticides

After food is harvested it is sometimes sprayed with pesticides to protect it from spoilage caused by bugs and moulds. Supporters of food irradiation claim that their process is safer than chemical pest control products, while ensuring an even longer shelf-life. Those on the other side of the debate point out that although irradiation could eliminate the use of some chemicals, the majority of pesticides are applied before harvest, while the plants are still growing. Irradiation will do nothing to cut down on the use of pesticides by farmers. And no one knows what effects irradiation will have on the residues of these chemicals. As well, there is evidence that irradiated products are more

TABLE 13 Irradiation and Additives

Some additives which it is claimed irradiation might replace		Some additives which might be needed to reduce undesirable effects of radiation	
sodium nitrite	m,c	sodium nitrite	m,c
sulphur dioxide	c	sodium sulphite	i
benzoic acid	a,i	ascorbic acid	–
propylene glycol	c	butylated hydroxyanisole (BHA)	c
chlorine	i	butylated hydroxytoluene (BHT)	c
chlorine dioxide	i	potassium bromate	i
sodium proprionate	a	sodium tripolyphosphate	i
ethylene oxide	c,i,t,m	sodium chloride (salt)	–
propylene oxide	i	niacin (for vitamin B_6)	–
hydrogen peroxide	i	sodium ascorbate (for vitamin C)	

i = irritant

c = carcinogen (known or suspected cancer causing agent in animals and/or humans)

m = mutagen (capable of causing mutations)

t = teratogen (capable of causing damage to developing foetus)

a = allergen (capable of causing allergic reactions)

Adapted from: Webb, Tony, Tim Lang and Kathleen Tucker. *Food Irradiation: Who Wants It?* Wellingborough, Northhamptonshire: Thorsons Publishing Group, 1987, p. 81.

Note: Only those additives allowed to be used in Canada have been listed.

susceptible to mould and fungal diseases. This would increase, rather than decrease, the amount of fungicides needed to protect food after it has been harvested.

Food Poisoning

Supporters of food irradiation say that it can be used to reduce outbreaks of salmonella and other types of food-poisoning in humans who eat poultry and seafood. Critics say that these outbreaks could be, and should be, reduced just as effectively by cleaning up food-processing plants and educating consumers on how to best handle, store and cook these foods. They fear that the use of irradiation could give manufacturers and consumers a false sense of security and lead them to handle food less carefully.

Other Concerns

When food goes bad it often smells unpleasant or looks mouldy, due to the presence of bacteria. Some of these bacteria produce toxic by-products that make people ill and cause, for example, food-poisoning. People are often able to avoid food-poisoning because other clues, such as an offensive odour or mould, tell them that the food is not fit for consumption. Irradiating foods will kill the bacteria, but will not eliminate the toxins that they produced. This means that food may look and smell perfectly fine and yet still be contaminated with dangerous toxins. Signs that would otherwise warn us that we should not eat the food will have been prevented from developing.

Some evidence suggests that although irradiation initially suppresses mould growth, foods may actually be more susceptible to becoming infected again. Certain foods such as nuts and grains may be more likely to support the growth of aflatoxins. Produced by natural moulds, aflatoxins are potent cancer-causing agents. Health and Welfare Canada has said that the

STUDIES AND FURTHER STUDIES

"Soviet studies indicate that rats fed irradiated food suffered higher rates of kidney and testicular damage. Canadian studies indicate that laboratory animals fed irradiated food developed an extra set of chromosomes. In Hungary, 1,223 studies on food irradiation safety have been conducted and not one of them supports the safety of irradiated food. And in West Germany food irradiation is banned because of studies indicating the possibility of mutations, reduced fertility, metabolic disturbances, decreased growth rate, reduced resistance to diseases, changes in organ weight, and cancer."

Excerpted from Steinman, David. *Diet For A Poisoned Planet.* New York: Harmony Books, 1990, p. 197.

research suggesting this connection between irradiation and increased aflatoxin production is not relevant because the study was conducted under conditions not found in commercial practice. They cite other studies that suggest irradiation can actually destroy aflatoxin in food. Another concern regarding food irradiation is that some bacteria may be mutated in the process of irradiation and become more toxic.

Even if food is sterile after it has been irradiated, there is nothing to keep it from becoming contaminated again later during food-processing, storage or distribution. People may wrongly assume that once a food is irradiated it is immune to any further infection.

World Hunger

Supporters of food irradiation suggest that it might help efforts to feed starving people in developing countries; irradiation could reduce the spoilage rate of the world's food supply from its estimated current rate of 25 to 30 percent. First, let's make one thing perfectly clear: There is no shortage of food in the world. In fact, there is enough food to feed adequately every single person on the planet today. In Canada and the United States there are tonnes and tonnes of surplus grain stored in silos. Europe has more surplus food than it can store. Yet millions of people still starve to death every year – not because there is a shortage of food, but because they have neither the money to buy food nor the resources to grow their own. The best agricultural lands in developing countries are often owned by corporations based in industrialized countries. The "cash crops" grown on these lands, such as coffee, bananas and sugar cane, are exported to developed countries like Canada where there is already an abundance of food. Hunger is a matter of food distribution, economics and, mostly, politics. Although food irradiation could facilitate distribution, it will do nothing to change the occurrence of starvation.

FOOD SAFETY

Nutrition

There is an abundance of evidence to suggest that some of vitamins A, C, E, K, B_1, B_2, B_3, B_6, B_{12}, carotene and thiamine are lost during irradiation. Although the losses vary from one food to another, it is generally agreed that vitamin losses increase as the irradiation dose increases.

Supporters of food irradiation acknowledge the process causes nutritional losses. However, they say that the losses are the same as those that result from other food processing methods. They go on to suggest that, in any case, such losses do not pose a problem for people eating a varied, generally nutritious diet. Others suggest that irradiated foods should be fortified with vitamins, or that people should take vitamin supplements. These arguments

VINDICATOR

The name could easily be mistaken for a character from a Saturday morning cartoon, but Vindicator is hardly child's play. Located in Mulberry, Florida, Vindicator is the first commercial food-irradiation plant in North America. It began operation in January 1992 by irradiating 1,000 pints of strawberries in order to prolong their shelf-life, kill insects and delay ripening. These were shipped to grocery stores in Florida, Virginia and Chicago. The facility is designed to irradiate 600 million tons of food each year. For Sam Whitney, owner of the plant, poultry is the main target.

Should this concern us as Canadians? Absolutely. Canadian dollars have been supporting this plant. Nordion International Inc. of Kanata, Ontario, supplied the irradiation equipment, some of the financing and design plans, and continues to supply the radioactive material – cobalt-60 – to Vindicator. Many observers anticipate that Vindicator will soon ask Ottawa to let them ship irradiated produce into Canada. Canadians are major customers for U.S. strawberries and grapefruit, both of which will be irradiated at the facility. If this happens, radiation equipment and radioactive material would leave Canada for the United States to irradiate food which would then be shipped backed for Canadian consumption.

But negative consumer reaction to irradiated food may ultimately dash the hopes of those who support it. According to Food & Water Incorporated (New York, NY), the first test market sale of irradiated strawberries at a store in Florida was not successful. While the press were gathered at the store, the fruit was selling quickly. But apparently many of the buyers were Vindicator stockholders. After the media left, sales dropped dramatically. *The Food Activist* (April 1992) reported: "During the next week, the price of irradiated berries had to be dropped from $1.89 to $.99 and 20 flats of unsold berries were shipped to a Miami wholesaler. The wholesaler then resold the irradiated berries to several food services where the berries do not need to be labelled for the consumer." Unfortunately only the first part of the story made it into the papers. Headlines such as "Irradiated Berries Find Lots of Buyers Despite Pickets" appeared in papers all over the United States.

seem to conflict with the suggestion that irradiating foods will help feed undernourished people in developing countries who often rely on a very few staple foods. (There are also fears that malnourished people may be more at risk for chromosomal changes than healthier individuals.) Donald Louria, in *The Bulletin of the Atomic Scientists* (September 1990) says, "In less developed countries, reducing the food supply's nutritional value would seem to raise a major ethical question. Asking the world's 800 million malnourished and 2 billion undernourished to make a possible trade-off between longer shelf-life and less nutrition seems harsh, particularly before more complete information on the nutritional value of irradiated foods is available." Critics of food irradiation argue that since irradiation is typically used in combination with other processing methods, such as cooking, losses in nutritional value may be substantial.

Chemical Changes in Food

One thing we know for sure is that irradiation does not make food radioactive. Using electronic or machine sources to produce X-rays or electron beams, it *is* possible to induce radioactivity in foods if such sources are operated at high energy levels. However, the energy levels of gamma rays given off by Cobalt-60 and Cesium-137 are stable and are too low to induce radioactivity.

Something else we know is that chemical changes occur when foods are processed. One of these changes involves the production of substances called "free radicals." In foods, free radicals can cause human cancer as well as premature aging. Both cooking and irradiation produce free radicals in food, though irradiation seems to create more of these substances than do conventional cooking methods. A 1989 report from the International Organization of Consumer Unions cites the work of Australian physicists at Melbourne University who found

EXPORTING IRRADIATION

For Atomic Energy of Canada Limited (AECL) the broad clearance for food irradiation in Canada will not only help to clear the way for domestic sales, but will be especially advantageous for export market sales. If Canada uses the technique, or at least has regulations for its use on foods that we may potentially purchase from developing countries, it will certainly help sales agents from Canada's irradiation industry sell their facilities around the world.

Adapted from: Kock, Irene. *The Facts About Food Irradiation*. Oshawa, Ontario: Nuclear Awareness Project, 1988, pp. 21-22.

that, in foods exposed to gamma radiation, the levels of free radicals increased between three and fifty times. Free radicals react with components of the food and new substances, called radiolytic products, develop. Many of these compounds are unique to the irradiation process and are thus known as unique radiolytic products (URPs). These completely new substances have not been tested to determine if they cause cancer, birth defects or chromosomal mutations. Their safety is virtually unknown.

Changes in the Body

Several studies have documented various adverse effects associated with food irradiation. Most tests have been done on animals, with the following results observed: earlier death; fewer offspring; increased incidence of stillbirths; lower birth weight; retarded growth; chromosomal damage; tumours; cataracts; and damage to bones, liver, spleen, kidneys, testicles and ovaries. No one knows what the effects might be of eating irradiated foods over a life-time. There may even be effects that won't show up until later generations.

Risks to Workers and the Environment

If we begin to irradiate large quantities of our food supply, more irradiation facilities will have to be built. There are already some 140 facilities worldwide. And more facilities will mean that more radioactive cobalt and cesium will be handled and transported. The dangers involved in this make a lot of people nervous. Workers in irradiation plants are at risk for developing cancer, genetic damage and weakening of the immune system if they are exposed.

Irradiation supporters refer to the fact that for twenty-five years medical supplies in North America have been sterilized in irradiation plants without incident. They feel that these plants have a good safety record, better than that of the chemical, oil and coal industries. Food irradiation opponents say that claiming to be better than other destructive industries does not prove safety. And critics refer to examples such as New Jersey, which has the highest concentration of irradiation plants anywhere in the United States. There has been at least one incident of environmental contamination at almost every plant in that state.

IS IT NEEDED?

Spices are currently the only irradiated foods in Canada and may well be the most commonly irradiated food item worldwide. Irradiation is used to control the bacteria and insects that commonly invade spices. Spices were traditionally disinfected with chemicals such as ethylene dibromide and ethylene oxide. Residues of these chemicals remaining on the food have been linked to the development of cancer, birth defects and chromosomal damage. Some countries have banned their use. Refrigeration controls insect contamination of spices but is an expensive solution. Irradiation seems to be an answer for dealing with both bacteria and insects. But there is an alternative. Spice companies have started using rapid steam heating under pressure to stop contamination. And it seems to be working.

In most other cases food irradiation is also unnecessary. It is often used like a prescription for an ill patient, providing a quick after-the-fact "cure" to a problem rather than focusing on preventing the problem (diseases and contamination) from developing in the first place. We already know of safe, effective ways to preserve, store and handle foods. We just need to use them properly. The food industry must spend the time and money necessary to train their workers in keeping food processing operations clean. And consumers must learn how to handle, store and cook foods such as poultry correctly. If so, salmonella poisoning could be virtually eliminated – without the use of food irradiation. Indeed, that is exactly what has been done in Sweden.

Although pesticides are often applied to produce after harvest in order to prevent disease and rotting, they are not the only option. And neither is irradiation. Fruits and vegetables may be stored at low temperatures, sometimes in the presence of carbon dioxide, to retard spoilage and prevent infestation. An alternative method is to treat produce with a steam heat process. In Hawaii, officials rejected federal funds offered to build an irradiation facility for processing papaya and instead chose to use non-chemical treatments such as dry and steam heat and double hot water dips.

Another barrier to food irradiation is the cost involved. It is an expensive process. For relatively inexpensive foods, such as wheat, potatoes and onions, it just isn't worth it. For other, more expensive items, such as poultry, fish or tropical fruits, it may be. But the few extra pennies that you'll have to pay for irradiated food doesn't include the health care or clean-up costs that may be involved should an accident occur.

THE LAST WORD

Pollution Probe has concluded that the irradiation of foods should be opposed until:

- legislation is in place that all foods, *any* portion of which have been irradiated, are clearly labelled as such
- it is clearly shown that food irradiation is necessary and that alternative methods of preventing food disease and contamination are used before turning to quick-fix cures
- nutritional losses due to irradiation are more carefully studied
- potential dangers to workers, the public and the environment are carefully evaluated
- restaurants are required to identify irradiated ingredients in their meals.

PART THREE:

ADDITIVES: HOW TO AVOID THEM

13

EATING OUT

THE NEXT TIME you go grocery shopping you'll be reading labels and looking out for suspect food additives. But what about when you eat out? Restaurants often have menus that vary from day to day. It may be too much to expect a restaurant to provide a complete list of ingredients for each dish they serve, particularly if that dish is only served for one day or a few weeks. Fast-food restaurants are another story, however.

Much of the success of fast-food restaurants is due to their predictability and consistency. It doesn't seem too much to ask for a list of ingredients for a food that is prepared and served almost exactly the same way thousands of times a day every day. And yet the paper wrapped around a hamburger or the cup holding a shake gives no list of the ingredients in that product. Consumers wishing to avoid particular ingredients, including food additives, are almost at a complete loss. Those with food allergies are particularly vulnerable.

We thought it would be useful, then, to provide a listing of the additives that can be found in each menu item offered by the major fast-food chains. We chose those restaurants that have outlets in at least 10 of the 12 Canadian provinces and territories and asked them to send us a list of ingredients for all of their products. The responses we received were quite varied. McDonald's, for example, has such a list available in all of their restaurants. Customers only have to ask for it.

Other restaurants assured us that they were currently working on putting such a list together, but that it wouldn't be available for a few months. When we called back a few months later it still wasn't available. We were assured that they were still working on it but it wouldn't be ready for a few more months.

Some restaurants were shocked that we would make such a request. "You want to know what is in *all* of our products? But there are so many of them!" These chains didn't have any such list available or any plans to have one available in the future. We were told that they would try to put one together for us – but that it would take a considerable amount of work and they couldn't promise when it would be ready.

Despite numerous requests and patient waiting none of the following restaurants provided us with a list of ingredients in their food products:

- A & W Food Services of Canada Ltd.
- Arby's (Canada) Inc.
- Domino's Pizza of Canada Inc.
- Kentucky Fried Chicken
- Mr. Submarine Ltd.
- Pizza Hut

How do you feel about eating at restaurants that can't, or won't, tell you what is in the food they are serving you?

The following restaurants did provide us with a complete list of ingredients for all of their food products. Please note: **We have listed only the food additives that are in the food – *not* all of the ingredients.** Also note that these lists are subject to change over time and from area to area. Some restaurants order the same products, hamburger buns for example, from a number of suppliers. Suppliers may vary in their formulations for a certain food item. For this reason, if you are particularly concerned about a food ingredient, especially in the case of food allergies, ask to see the label on the package that the product came in before placing your order. The restaurant always has this information available and is required to provide it to you.

TABLE 14 Additives in McDonald's Foods*

FOOD TYPE	FOOD ITEM	ADDITIVES
Meat	Canadian back bacon	sodium tripolyphosphate, sodium erythorbate, sodium nitrite
	Circular bacon	smoke flavour, sodium tripolyphosphate, sodium erythorbate, sodium nitrite
	Strips (side) bacon	sodium tripolyphosphate, sodium erythorbate, sodium nitrite
	Regular Hamburger – meat only	–
	Quarter Pounder – meat only	–
	Big Mac – meat only	–
	McLean Deluxe – meat only	flavour (yeast extract, maltodextrine, citric acid)
	Sausage Patty – meat only	–
	Frozen sausage spice blend	calcium silicate (processing aid)
Beverages	Orange	citric acid, natural flavours, sodium benzoate, colour
	Root beer	caramel colour, natural flavours, sodium benzoate, colour
	Coffee	–
	Hot chocolate	disodium phosphate, lecithin, artificial flavour

FOOD TYPE	FOOD ITEM	ADDITIVES
Beverages *cont'd*	Orange juice	–
	Apple juice	–
	Lowfat milkshake product	microcrystalline cellulose, guar gum, cellulose gum, carrageenan, artificial flavour
	Chocolate milkshake syrup	sodium benzoate, citric acid, artificial flavour
	Strawberry milkshake syrup	citric acid, sodium benzoate, propylene glycol, artificial colour, artificial flavour, lactic acid
	Vanilla milkshake syrup	propylene glycol, citric acid, caramel colour, sodium benzoate, artificial flavour
	Marshmallows (for hot chocolate)	tetrasodium pyrophosphate, artificial flavour
Bread products	Buns	calcium sulphate, may contain any or all of the following in varying proportions: sodium stearoyl-2-lactylate, mono- and diglycerides, ammonium chloride, potassium bromate, potassium iodate, ascorbic acid, calcium propionate, dicalcium phosphate, calcium carbonate, calcium iodate, protease, calcium peroxide, aziodicarbonamide, diammonium phosphate, amylase, diacetyl tartaric acid esters of mono and diglycerides, sorbic acid

FOOD TYPE	FOOD ITEM	ADDITIVES
Bread products *cont'd*	English muffins	calcium propionate, calcium sulphate, and may contain the following in various proportions: monocalcium phosphate, L-cysteine hydrochloride, ammonium chloride, ascorbic acid, protease enzyme, diacetyl tartaric esters of mono- and diglycerides, potassium sorbate, potassium bromate, calcium iodate, calcium carbonate
	French bread	calcium sulfate, calcium propionate, potassium bromate, and may contain any or all of the following in various proportions: ammonium chloride, ammonium phosphate, magnesium carbonate, ascorbic acid, L-cysteine chloro-hydrate, mono- and diglycerides, azodicarbonamide, calcium iodate
	Lowfat no-cholesterol apple bran muffin	mono- and diglycerides, butter flavour, propylene glycol monostearate, sodium stearoyl lactylate and diacetyl tartaric acids, esters of mono- and diglycerides, BHA, citric acid
	Hot cakes	sodium aluminum phosphate, dried egg white (triethyl citrate, may contain: lipase, sodium lauryl sulphate, citric acid, lactic acid, sodium carbonate), polysorbate 60
Cereals	Cheerios[1]	calcium carbonate, colour, trisodium phosphate

[1] General Mills product

FOOD TYPE	FOOD ITEM	ADDITIVES
Cereals *cont'd*	Honey Nut Cheerios[1]	trisodium phophate
Cheese	Processed slices	sodium phosphate and/or sodium citrate, rennet and/or microbial enzyme and/or pepsin, calcium chloride, sorbic acid, natural colour, and may contain potassium sorbate, citric acid, carboxymethyl cellulose, starch and/or lecithin
Poultry	Chicken McNuggets – meat only	cooked in 100% vegetable oil (see Shortening blends)
	Chicken McNuggets – tempura batter	modified starch, leavening (sodium bicarbonate and sodium aluminum phosphate), spice
	Chicken McNuggets – first batter	leavening (sodium bicarbonate and sodium aluminum phosphate)
	Chicken McNuggets – breading	spice
Desserts	Cookies – Chocolaty Chip	sweet chocolate drops (soya lecithin, vanillin), cocoa drops (soya lecithin, sorbitan monostearate, polysorbate 60, vanillin), flavour modified butter, leavening (sodium bicarbonate, sodium acid pyrophosphate, monocalcium phosphate), soya lecithin, artificial flavour
	Cookies – McDonaldland	leavening (sodium bicarbonate, sodium acid pyrophosphate, monocalcium phosphate) soya lecithin, natural flavour

[1] General Mills product

FOOD TYPE	FOOD ITEM	ADDITIVES
Desserts *cont'd*	Danish, apple	bleached dried apples (contain sulfur dioxide), mono- and diglycerides, citric acid, potassium sorbate, cinnamon, sodium stearoyl-2-lactylate, calcium carbonate, calcium sulphate, agar, locust bean gum, titanium dioxide, BHA, sodium benzoate, sodium propionate, lecithin, artificial flavour, propyl gallate, xanthan gum, turmeric, paprika
	Danish, raspberry	mono- and diglycerides, citric acid, potassium sorbate, sodium stearoyl-2-lactylate, calcium carbonate, calcium sulphate, locust bean gum, titanium dioxide, BHA, algin, sodium benzoate, sodium propionate, artificial flavour
	Figure "8" Danish – pastry portion	dough conditioner (L-cysteine hydrochloride)
	Figure "8" Danish – apple filling	modified food starch, citric acid, sodium benzoate (as a preservative)
	Figure "8" Danish – cherry filling	modified food starch, citric acid, artificial colour, sodium benzoate, potassium sorbate
	Figure "8" Danish – cheese	–
	Figure "8" Danish – egg wash	–

FOOD TYPE	FOOD ITEM	ADDITIVES
Desserts *cont'd*	Birthday cake – chocolate	vegetable oil (mono- and diglycerides, BHA, BHT, citric acid), baking powder (sodium bicarbonate, sodium pyrophosphate, monocalcium phosphate, calcium lactate), sodium bicarbonate, and may contain mono- and diglycerides, xanthan gum, natural colour, sorbitan monostearate, polysorbate 60
	Birthday cake – yellow	vegetable oil (BHA, BHT), baking powder (sodium bicarbonate, sodium pyrophosphate, monocalcium phosphate, calcium lactate), artificial flavour, sorbitan monostearate, polysorbate 60, and may contain mono- and diglycerides, xanthan gum, natural colour
	Birthday cake – icing	vegetable oil (mono- and diglycerides, BHA, BHT, citric acid), artificial flavour, colour, may contain xanthan gum, polysorbate 60
	Frozen yogurt, lowfat	microcrystalline cellulose, mono- and diglycerides, guar gum, cellulose gum, artificial flavour, carrageenan, colour
	Frozen yogurt, lowfat chocolate	microcrystalline cellulose, mono- and diglycerides, guar gum, cellulose gum, artificial flavour, carrageenan, colour
	Apple pie filling	modified corn starch, spices, citric acid, sodium benzoate, calcium chloride, erythorbic acid

FOOD TYPE	FOOD ITEM	ADDITIVES
Desserts *cont'd*	Apple pie spice blend unit	lactose, microcrystalline cellulose, polysorbate 80, natural colour (annatto), artificial flavour (ethyl vanillin), tricalcium phosphate
	Blueberry pie filling	modified corn starch, natural flavour, citric acid, spices
	Cherry pie filling	modified corn starch, citric acid, natural and artificial flavour, colour
	Raspberry pie filling	modified corn starch, natural flavour, artificial colour
	Strawberry-rhubarb pie filling	modified corn starch, natural flavour (strawberry flavour, propylene glycol), citric acid, artificial colour
	Pie pastry	guar gum, mono- and diglycerides, sodium carbonate, L-cysteine hydrochloride, citric acid
	Cones	soya lecithin, sodium bicarbonate, artificial flavour, artificial colours, sodium benzoate, may also contain citric acid, magnesium carbonate, protease, cream of tartar
Desserts – sundae toppings	Hot caramel	pectin, disodium phosphate, artificial flavour
	Hot fudge	sodium propionate, sodium phosphate, potassium sorbate, artificial flavour

FOOD TYPE	FOOD ITEM	ADDITIVES
Desserts – sundae toppings *cont'd*	Strawberry	citric acid, locust bean gum, pectin, natural and artificial flavour, sodium benzoate, calcium phosphate, artificial colour
Fish	Filet-O-Fish portion	modified corn starch, cellulose gum, spice, paprika and turmeric extract colour, natural flavouring, may contain sodium phosphates (to retain freshness)
	Filet-O-Fish batter	modified corn starch, cellulose gum, spice, manufacturing aids (propylene glycol, calcium silicate)
	Filet-O-Fish Krusto breading	–
Potato Products	French fries	may contain sodium acid pyrophosphate
	Hash browns	sodium acid pyrophosphate (to promote colour retention), spice
Condiments and toppings	Hot cake syrup	natural and artificial maple flavour, potassium sorbate as a preservative, artificial colour
	Ketchup	natural flavourings
	Margarine	sorbitan tristearate, vegetable lecithin, mono- and diglycerides, sodium benzoate, potassium carbonate, coloured with beta-carotene, artificial flavour

FOOD TYPE	FOOD ITEM	ADDITIVES
Condiments and toppings *cont'd*	Butter, whipped	colour
	Mustard, prepared	–
	Onions, dehydrated	–
	Dill pickle slices	calcium chloride, sodium benzoate, potassium sorbate, spices, alum, polysorbate 80, turmeric
	Granulated nuts	–
Spreads	Orange marmalade	natural fruit pectin, citric acid
	Peanut butter	–
	Grape jam with pectin	malic acid, pectin, sodium citrate
	Strawberry preserves with pectin	pectin, citric acid, sodium citrate
Poultry	McChicken – meat only	flavouring, cooked in 100% vegetable oil (see Shortening blends)
	McChicken – flavouring	hydrolyzed plant protein, flavour, flavour enhancers (disodium inosinate, disodium guanylate)
	McChicken – first batter	modified starch, spice, sodium aluminum phosphate, sodium bicarbonate

FOOD TYPE	FOOD ITEM	ADDITIVES
Poultry *cont'd*	McChicken – first breader	modified starch, sodium bicarbonate, sodium aluminum phosphate, sodium bicarbonate
	McChicken – second batter	modified starch, spice, sodium aluminum phosphate, sodium bicarbonate
	McChicken – second breader	modified starch, sodium aluminum silicate, spice, sodium aluminum phosphate, sodium bicarbonate
Sauces	McChicken sauce	modified corn starch, spices, xanthan gum, potassium sorbate, beta-carotene, calcium disodium EDTA
	Tartar sauce (Filet-O-Fish)	dill pickle relish (potassium sorbate, spice), modified corn starch, spices, xanthan gum, potassium sorbate, calcium disodium EDTA
	Barbecue sauce (Chicken McNuggets)	spices and seasonings (hydrolyzed plant protein, disodium inosinate, disodium guanylate, tricalcium phosphate, artificial flavour, caramel), modified corn starch, vegetable oil (may contain BHA, BHT and mono- and diglycerides), xanthan gum, potassium sorbate, sodium benzoate
	Honey sauce (Chicken McNuggets)	–

FOOD TYPE	FOOD ITEM	ADDITIVES
Sauces *cont'd*	Hot mustard sauce (Chicken McNuggets)	prepared mustard (spices, artificial colour), vegetable oil (may contain BHA, BHT and mono- and diglycerides), spices, modified corn starch, xanthan gum, sodium benzoate, caramel and natural colour, calcium disodium EDTA
	Sweet and sour sauce (Chicken McNuggets)	spices and seasonings, modified corn starch, vegetable oil (may contain BHA, BHT and mono- and diglycerides), xanthan gum, potassium sorbate, natural flavour, malic acid, carboxymethyl cellulose, caramel colour
	Big Mac sauce	pickle relish (potassium sorbate, xanthan gum, spices), modified corn starch, spices, potassium sorbate, xanthan gum, onion flavour, hydrolyzed vegetable protein, calcium disodium EDTA
Salad Dressings	French	spices, xanthan gum, propylene glycol alginate, calcium disodium EDTA, natural colour (oleoresin paprika)
	House	creaming agent (sodium caseinate, mono- and diglycerides, sodium citrate, dipotassium phosphate, carrageenan, sodium stearoyl-2-lactylate), spices, natural flavour, sodium benzoate, xanthan gum, calcium disodium EDTA

FOOD TYPE	FOOD ITEM	ADDITIVES
Salad dressings *cont'd*	Thousand Island	sweet relish (spices, spice flavours, xanthan gum, sodium benzoate), sour cream powder (citric acid, BHA), lactic acid, spices, xanthan gum, calcium disodium EDTA
	Vinaigrette	spices, propylene glycol alginate, xanthan gum, potassium sorbate, natural colour, calcium disodium EDTA
Salad Toppings	Julienne ham	sodium tripolyphosphate, sodium erythorbate, sodium nitrite
	Julienne turkey	carrageenan
Shortening blends	For french fries and hash browns	monoglyceride citrate, propyl gallate, propylene glycol
	100% vegetable oil for Chicken McNuggets, Filet-O-Fish, Hot Pie Desserts, McChicken	monoglyceride citrate, propyl gallate added to protect flavour, propylene glycol

* Adapted from the booklet *McDonald's Food: The Ingredient Facts . . .* provided by McDonald's. According to the pamphlet, "A product's actual ingredients are subject to some variation depending on the local supplier, the region of the country and the season of the year. Therefore, the ingredient lists in this book may not necessarily be exactly identical to the product served in a particular restaurant. However, any difference would be slight, since all products served in McDonald's – and provided by our suppliers – meet McDonald's strict specifications and high standards of quality.

"Also, McDonald's, from time to time, reviews the recipes of its menu items and makes changes in these recipes in an effort to bring you the highest quality, best-tasting food. The ingredient listings in this booklet are effective as of May 1, 1991. McDonald's will publish updated versions of this booklet periodically."

TABLE 15 Additives in Burger King Foods*

FOOD TYPE	FOOD ITEM	ADDITIVES
Buns	Whopper	sodium stearoyl-2-lactylate, polyoxyethlene (8) stearate, calcium propionate, calcium sulphate, ammonium chloride, calcium carbonate, calcium iodate
	Burger	*see* Whopper bun
	Specialty bun	*see* Whopper bun
Bread products	Croissant	mono- and diglycerides, calcium propionate, yeast nutrient (calcium sulphate, ammonium sulphate), annatto, turmeric and paprika extracts, sodium phosphate, lecithin, carrageenan
	Oat bran bun	calcium proprionate, lecithin, sodium stearoyl-2-lactylate, polyoxyethylene (8) stearate, calcium sulphate, ammonium chloride, calcium carbonate, calcium iodate, artificial flavour, caramel colour
	Bagel	bromated flour (potassium bromate, azodicarbonamide), artificial flavour, sodium propionate, fumaric acid
	French toast sticks	contains 2% or less of the following: dough conditioners (may contain one or more of the following: mono- and di-glycerides, ethoxylated mono- and diglycerides, calcium

FOOD TYPE	FOOD ITEM	ADDITIVES
	French toast sticks *cont'd*	and sodium stearoyl lactylates, calcium peroxide), yeast nutrients (monocalcium phosphate, calcium sulphate, ammoniun sulphate), calcium propionate, turmeric and paprika extractives. Batter and breading: contains 2% or less of the following: modified corn starch, lecithin, gum arabic, leavening (monocalcium phosphate, sodium bicarbonate), glycerine, natural and artificial flavour, polysorbate 80, carrageenan
Meat	Breakfast sausage	seasoning
	Bacon	sodium phosphate, sodium erythorbate, sodium nitrite
	Ham, diced	sodium phosphate, carrageenan, sodium erythorbate, sodium nitrite
	Ham, sliced	sodium phosphate, carrageenan, smoke flavour, sodium erythorbate, sodium nitrite
	Turkey, diced	modified corn starch
Fish	Ocean Catch fish filet	skinless and boneless cod filets (may contain sodium phosphate), modified starch, spices. Prepared in 100% vegetable shortening.

FOOD TYPE	FOOD ITEM	ADDITIVES
Chicken	Chicken Cutlette	Batter: modified starch. Breading: spice, artificial flavour. Prepared in 100% vegetable shortening.
	BK Broiler Chicken Cutlette	natural flavourings, spice, modified corn starch, MSG, calcium chloride. Glazed with: natural flavour, modified starch, methylcellulose, xanthan gum, spice.
	Chicken Tenders	spices. Prepared in 100% vegetable shortening.
	Diced	modified food starch, MSG, calcium chloride
Cheese		sodium citrate, sodium phosphate, sorbic acid, lecithin, acetic acid, citric acid, rennet, pepsin, colour, calcium chloride
Condiments and toppings	Pickles	spices, polysorbate 80, alum, sodium benzoate, potassium sorbate
	Ketchup	spices
	Mayonnaise	spice, calcium disodium EDTA
	Mustard	turmeric, spices
Sauces	Barbecue sauce	flavours, colour, spices, seasonings

FOOD TYPE	FOOD ITEM	ADDITIVES
Sauces *cont'd*	Sweet and sour sauce	modified corn starch, hydrolyzed vegetable protein, flavour, Worchestershire sauce (spices, flavour, colour), sodium benzoate, spices
	Honey sauce	caramel colouring
	BK broiler sauce	dijon mustard (tartaric acid, citric acid, spices), spices, Worchestershire sauce (spices, flavour, colour), MSG, xanthan gum, mustard flavour, potassium sorbate, calcium disodium EDTA
	Tartar sauce	relish (modified starch, guar gum, sodium benzoate, spices, polysorbate 80, colour), acetic acid, spices, calcium disodium EDTA
Potato Products	French fries	disodium dihydrogen pyrophosphate. Prepared in 100% vegetable shortening.
	Tater Tenders	natural flavouring, disodium dihydrogen pyrophosphate. Prepared in 100% vegetable shortening.
Mini-Onion Rings		modified corn starch, natural onion flavour, calcium chloride, methylcellulose, guar gum, sodium alginate, sodium bicarbonate, glucono delta lactone, sodium

FOOD TYPE	FOOD ITEM	ADDITIVES
Mini-Onion Rings *cont'd*		tripolyphosphate, sodium acid pyrophosphate, sodium aluminum phosphate, hydroyxypropyl methylcellulose, sorbitol. Prepared in 100% vegetable shortening.
Egg Mix		grilled in 100% vegetable shortening
Salad Dressings	French	paprika, spices, natural flavour, xanthan gum
	Thousand Island	sweet pickle relish (xanthan gum), spices, xanthan gum, natural flavour
	Ranch	natural flavours, spices, xanthan gum
	Light oil and vinegar	spices, xanthan gum, turmeric
Beverages	Apple juice	ascorbic acid
	Hot chocolate	disodium phosphate, potassium sorbate, lecithin, artificial flavours
	Vanilla shake	cellulose gum, mono- and diglycerides, guar gum, sodium phosphate, sodium citrate, sodium carbonate, carrageenan
	Chocolate shake	same as vanilla shake. The following syrup is added: modified corn starch, potassium sorbate, xanthan gum, citric acid, artificial flavour

FOOD TYPE	FOOD ITEM	ADDITIVES
Beverages *cont'd*	Strawberry shake	same as vanilla shake. The following syrup is added: citric acid, sodium benzoate, artificial flavour, colour ethyl maltol
Desserts	Apple pie	modified starch, potassium sorbate, cinnamon, lecithin, sodium phosphate, carrageenan, extracts of turmeric and paprika
	Cherry pie	modified starch, potassium sorbate, sodium phosphate, lecithin, carrageenan, extracts of turmeric and paprika
	Cookies and Cream Stir-Ups	same as vanilla shake. The following topping is added: soya lecithin, artificial flavours, ethyl maltol
	Apple Cinnamon Swirl Danish	apples (natural and artificial apple flavourings), mono- and diglycerides, leavening agents (sodium bicarbonate, sodium acid pyrophosphate), natural and artificial flavourings, dough conditioners (sodium stearoyl-2-lactylate, calcium sulphate, ammonium sulphate)
Shortenings	Vegetable shortening (for all fried products)	dimethylpolysiloxane

* This information was adapted from a list of ingredients provided by Burger King which states, "Product data is current as of the date of publication. New product introductions or product changes may cause deviations from the information provided." Published 1989.

TABLE 16 Additives in Wendy's Foods*

FOOD TYPE	FOOD ITEM	ADDITIVES
Buns	Kaiser	calcium sulfate, sodium stearoyl lactylate, calcium stearoyl-2-lactylate, turmeric, paprika, sodium caseinate, monocalcium phosphate, potassium bromate, azodicarbonamide
	Sandwich	(as for kaiser buns)
Meat	Single burger	–
	Big Classic burger	–
	Bacon	sodium phosphate, sodium erythorbate, sodium nitrite
	Chili	acid base (citric acid, calcium chloride), seasoning (maltodextrin, spices, modified food starch, monosodium glutamate, silicon dioxide, xanthan gum, citric acid, oleoresin paprika, disodium guanylate, disodium inosinate, artificial flavouring), vegetable mix (calcium chloride, calcium disodium EDTA), chili beans (natural flavourings, calcium chloride)
	Country fried steak	sodium phosphate. Breaded with: modified corn starch, spices. Cooked in vegetable oil.

FOOD TYPE	FOOD ITEM	ADDITIVES
Poultry	Grilled chicken breast fillet	sodium phosphates. Breaded with: modified corn starch, spices. Battered with: modified corn starch, spices, leavening (sodium acid pyrophosphate, sodium bicarbonate, monocalcium phosphate). Cooked in vegetable oil.
	Crispy chicken nuggets	sodium tripolyphosphate, monosodium glutamate. Battered and breaded with: modified food starch, leavening (sodium bicarbonate, monocalcium phosphate, sodium acid pyrophosphate, sodium aluminum phosphate), spices, citric acid, xanthan gum
	Chicken breast fillet	sodium phosphates. Breaded with: modified corn starch, spices. Battered with: modified corn starch, spices, leavening (sodium acid pyrophosphate, sodium bicarbonate, monocalcium phosphate). Cooked in vegetable oil.
Fish	Fish fillet	modified corn starch, leavening (sodium bicarbonate), lecithin, monosodium glutamate, hydrolyzed plant protein, spice. Cooked in vegetable oil.
Cheese	American cheese slice	cheese (artificial colour), sodium citrate, sodium phosphate, acetic acid, sorbic acid, lecithin, artificial colour

FOOD TYPE	FOOD ITEM	ADDITIVES
Cheese *cont'd*	Cheddar cheese, shredded	powdered cellulose or microcrystalline cellulose, artificial colour
Condiments and toppings	Ketchup	spices, natural flavourings
	Honey mustard	spices, xanthan gum, sodium benzoate, potassium sorbate, calcium disodium EDTA
	Mayonnaise	calcium disodium EDTA
	Mustard	turmeric, spices, natural flavours
	Pickles, dill	alum, potassium sorbate, sodium benzoate, natural flavourings, turmeric, polysorbate 80
	Sweet mustard	modified food starch, spices, xanthan gum, turmeric, paprika
	Sour cream	locust bean gum, carrageenan, sodium citrate
Potato products	French fries	disodium dihydrogen pyrophosphate. Cooked in vegetable oil.
	Plain potato	–
Sauces	Tartar sauce	spices, turmeric

FOOD TYPE	FOOD ITEM	ADDITIVES
Sauce *cont'd*	Barbecue sauce	modified food starch, natural smoke flavouring, spices, sodium benzoate
	Honey	(*please see package*)
	Sweet and sour sauce	modified food starch, spices, sodium benzoate, artificial colour, calcium disodium EDTA
Salad bar	Applesauce, chunky	–
	Bacon bits	sodium phosphate, natural smoke flavouring, sodium erythorbate, sodium nitrite, BHA, BHT
	Breadsticks	–
	Cheddar chips	annatto, apocarotenal, natural flavour, citric acid
	Cheddar cheese, shredded (imitated)	natural flavour, sodium aluminum phosphate, powdered cellulose, lactic acid, modified food starch, sodium citrate, sodium phosphate, sorbic acid, acetic acid, artificial colour, guar gum
	Chicken salad	salad base: modified food starch, potassium sorbate, xanthan gum, natural flavour, calcium disodium EDTA
	Chow mein noodles	–

FOOD TYPE	FOOD ITEM	ADDITIVES
Salad bar *cont'd*	Cole slaw	salad dressing (spice, natural flavours, arabic, guar and xanthan gum, calcium disodium EDTA), sodium erythorbate, vegetable gum, sodium benzoate, potassium sorbate
	Cottage cheese	citric acid, guar gum, locust bean gum, mono- and diglycerides, carrageenan, calcium sulfate, potassium sorbate
	Croutons	monosodium glutamate
	Eggs	frozen or preserved in sodium benzoate, citric acid, potassium sorbate
	Garbanzo beans	calcium disodium EDTA
	Parmesan cheese	sodium silicoaluminate
	Parmesan cheese (imitation)	modified food starch, natural flavourings, sodium phosphate, potassium sorbate, acetic acid, monosodium glutamate
	Pasta salad	spice, natural flavour
	Pepperoni, sliced	spices, lactic acid, oleoresin of paprika, flavouring, sodium nitrite, BHA, BHT, citric acid
	Potato salad	spices, sodium benzoate, potassium sorbate, xanthan gum

FOOD TYPE	FOOD ITEM	ADDITIVES
Salad bar *cont'd*	Pudding, butterscotch	modified food starch, sodium stearoyl lactylate, artificial flavouring, disodium phosphate, artificial colouring (includes FD&C Yellow #5)
	Pudding, chocolate	modified food starch, sodium stearoyl lactylate, artificial colouring, artificial flavouring
	Seafood salad	calcium carbonate, modified food starch, natural flavours, artificial flavours, annatto, cochineal. Salad base: modified food starch, spice, potassium sorbate, xanthan gum, natural flavour, calcium disodium EDTA
	Sour topping	mono- and diglycerides, lactic acid, citric acid, acetic acid, locust bean gum, carrageenan, artificial flavour, artificial colour, potassium sorbate
	Three bean salad	natural flavouring, spice
	Tuna salad	modified food starch, spice, potassium sorbate, xanthan gum, natural flavour, calcium disodium EDTA
	Turkey ham	sodium phosphate, natural smoke flavouring, sodium erythorbate, sodium nitrate

FOOD TYPE	FOOD ITEM	ADDITIVES
Salad dressings	Blue cheese	xanthan gum, potassium sorbate, citric acid, natural flavourings, calcium disodium EDTA, artificial flavourings, oleoresin paprika
	Celery seed	spices, xanthan gum, sodium benzoate, potassium sorbate, caramel colour, natural flavourings, FD&C Yellow #5, FD&C Blue #1
	French	xanthan gum, spices, natural flavourings, ß-apo-8-carotenal, calcium disodium EDTA
	Sweet red French	xanthan gum, propylene alginate, potassium sorbate, natural flavourings, caramel colour, oleoresin paprika, calcium disodium EDTA
	Hidden Valley ranch	modified food starch, monosodium glutamate, natural flavours, xanthan gum, spices, citric acid, sorbic acid, calcium stearate, artificial flavour, calcium disodium EDTA
	Italian Caesar	lactic acid, polysorbate 80, natural flavourings, sorbic acid, monosodium glutamate, xanthan gum, sodium benzoate, hydrolyzed vegetable protein, citric acid

FOOD TYPE	FOOD ITEM	ADDITIVES
Salad dressings *cont'd*	Golden Italian	xanthan gum, spices, natural flavourings, sodium benzoate, potassium sorbate, caramel colour, oleoresin paprika, FD&C Yellow #5
	Salad oil	—
	Thousand Island	sorbic acid, xanthan gum, natural flavours, calcium disodium EDTA
	Wine vinegar	—
	Reduced calorie bacon and tomato	bacon granules (hydrolyzed vegetable protein, natural and artificial flavours, caramel, artificial colour), xanthan gum, propylene glycol alginate, natural flavourings, polysorbate 60, sodium benzoate, potassium sorbate, monosodium glutamate, artificial flavourings, calcium disodium EDTA
	Reduced calorie Italian	propylene glycol alginate, xanthan gum, sodium benzoate, potassium sorbate, FD&C Yellow #5, calcium disodium EDTA
Salads, prepared	Chef salad	white turkey meat contains sodium phosphate; see imitation cheese
	Garden salad	*see* imitation cheese
	Taco salad	*see* chili, imitation cheese and taco chips

FOOD TYPE	FOOD ITEM	ADDITIVES
Superbar – Mexican Fiesta	Cheese sauce	cheese solids (artificial colour), modified food starch, sodium caseinate, natural cheese flavour, monosodium glutamate, dipotassium phosphate, citric acid, spices, sodium silicoaluminate, mono- and diglycerides, carrageenan
	Picante sauce	citric acid
	Refried beans	artificial colours
	Rice, Spanish	modified food starch, hydrolyzed vegetable protein, spices (including paprika and turmeric), monosodium glutamate, natural and artificial flavourings, citric acid, tricalcium phosphate, extractives of paprika, disodium inosinate, disodium guanylate
	Taco chips	–
	Taco meat	spices, monosodium glutamate, citric acid
	Taco sauce	modified food starch, spices, citric acid
	Taco shells	–
	Tortilla, flour	sodium propionate

FOOD TYPE	FOOD ITEM	ADDITIVES
Superbar – pasta	Alfredo sauce	parmesan cheese blend (lactic acids), modified food starch, natural flavours, monosodium glutamate, tricalcium phosphate, xanthan gum, mono- and diglycerides, carrageenan, artificial colour (including FD&C Yellow #5 and FD&C Yellow #6)
	Cheese ravioli in spaghetti sauce	modified food starch, spaghetti sauce (modified food starch, monosodium glutamate, spices, hydrolyzed vegetable protein, citric acid, disodium inosinate, disodium guanylate, artificial flavour, artificial colour)
	Cheese tortellini in spaghetti sauce	–
	Fettucine	–
	Garlic toast	artificial flavours, vegetable lecithin, dimethyl silicone
	Pasta medley	–
	Rotini	–
	Spaghetti sauce	modified food starch, monosodium glutamate, spices, citric acid, natural beef flavour, lemon flavour, natural flavours

FOOD TYPE	FOOD ITEM	ADDITIVES
Superbar – pasta	Spaghetti meat sauce	modified food starch, monosodium glutamate, spices, hydrolyzed vegetable protein, citric acid, disodium inosinate, disodium guanylate, artificial flavour, artificial colour
Desserts	Frosty dairy dessert	guar gum, carboxymethyl cellulose, mono- and diglycerides, carrageenan, disodium phosphate, natural flavouring, artificial flavouring
	Chocolate chip cookie	chocolate chips (lecithin, artificial flavours), modified food starch, leavening (baking soda, aluminum phosphate)
Beverages	Cola	caramel colour, phosphoric acid, natural flavourings, caffeine
	Diet cola	caramel colour, phosphoric acid, potassium benzoate, natural flavourings, citric acid, caffeine, aspartame, dimethylpolysiloxane
	Lemon-lime	citric acid, sodium citrate, sodium benzoate
	Coffee	–
	Decaffeinated coffee	–
	Hot chocolate	sodium caseinate, cellulose gum, vanillin

FOOD TYPE	FOOD ITEM	ADDITIVES
Beverages *cont'd*	Lemonade	citric acid, calcium carbonate, sodium carbonate, potassium carbonate, trisodium citrate, natural lemon flavour, monoglycerides, dimethylpolysiloxane, FD&C Yellow #5, xanthan gum, propyl gallate, BHA
	Milk (chocolate or 2%)	(*please see package*)
	Tea (hot or iced)	–

* Wendy's Restaurants of Canada does not have a completed list of ingredients for their menu items. Instead they provided a list from their U.S. restaurants and suggested that the information would be comparable.

TABLE 17 Additives in Taco Bell Foods*

FOOD TYPE	FOOD ITEM	ADDITIVES
Sauces	Taco sauce	spices, xanthan gum, sodium benzoate, natural flavour
	Hot taco sauce	spices, xanthan gum, sodium benzoate, natural flavour
	Cheese sauce	"protected trade secret and therefore not disclosed"
	Red sauce	"protected trade secret and therefore not disclosed"
	Green sauce	"protected trade secret and therefore not disclosed"
	Chilito sauce	"protected trade secret and therefore not disclosed"
	Pizza sauce	"protected trade secret and therefore not disclosed"
	Pico de Gallo	"protected trade secret and therefore not disclosed"
	Pico sauce	"protected trade secret and therefore not disclosed"
Meat	Seasoned beef	"protected trade secret and therefore not disclosed"
Poultry	Chicken	"protected trade secret and therefore not disclosed"
Condiments and toppings	Salsa	calcium chloride, malic acid, capsicom solution
	Sour cream	gelatin, vegetable gum (carob bean gum, guar gum, locust bean gum, carrageenan), sodium citrate, glyceryl monostearate

FOOD TYPE	FOOD ITEM	ADDITIVES
Condiments and toppings *cont'd*	Olives	ferrous gluconate
	Guacamole	sodium alginate, xanthan gum, citric acid, erythorbic acid
Salad dressings	Ranch	natural flavours, xanthan gum, spices, citric acid, sorbic acid, calcium stearate, artificial flavour, calcium disodium EDTA
Cheese	Cheddar	calcium chloride, annatto colour
	Pepper Jack	sodium citrate cellulose, sodium phosphate, sorbic acid, lactic acid
Tortillas	Enchirito	dough conditioners (sodium stearoyl-2-lactylate), potassium sorbate
	Flour	dough conditioners (sodium stearoyl-2-lactylate), potassium sorbate
	Flour – fried	dough conditioners (sodium stearoyl-2-lactylate), potassium sorbate
	Heat pressed	monoglycerides, dough conditioners (L-cysteine, sodium stearoyl-2-lactylate) calcium propionate, potassium sorbate
Desserts	Cinnamon Twists	natural flavour
	Pinto beans	artificial colours

* Although we made our request to the Canadian head office, this information came from the U.S. branch. You will notice that Taco Bell refused to list some of the ingredients in their products stating these "are protected trade secrets and therefore not disclosed." The information is up-to-date as of March 1992.

TABLE 18 Additives in Arby's Foods*

FOOD TYPE	FOOD ITEM	ADDITIVES
Sandwich fillings and buns	Bacon	sodium phosphate, sodium erythorbate, sodium nitrite
	Base mix for rolls	dough conditioners (vegetable mono- and diglycerides, potassium bromate, ascorbic acid, L-cysteine, enzyme)
	Deli bun	base mix (see above), sodium stearoyl lactylate, colour, double spice
	Poppy-seed bun	base mix (see above), spice mix, potassium bromate, calcium propionate
	Onion bun	base mix (see above), spice mix, potassium bromate, calcium propionate
	Breaded chicken breast fillet	breaded with flour (spices, MSG, hydrolyzed vegetable protein), leavening (sodium bicarbonate, sodium aluminum phosphate, monocalcium phosphate)
	Breaded fish fillet	modified starch, natural flavours, leavening (sodium bicarbonate, sodium aluminum phosphate), MSG, cellulose gum, spice, guar gum

FOOD TYPE	FOOD ITEM	ADDITIVES
Sandwich fillings *cont'd*	Cheddar cheese sauce	aged Cheddar cheese (enzymes, annatto colour), modified starch, Monterey Jack cheese (enzymes), sodium phosphate, annatto colour, FD&C Yellow #6
	Ham	sodium phosphate, smoke flavour, sodium ascorbate (erythorbate), sodium nitrite
	Mozzarella cheese	sodium proprionate, natural flavour
	Pasteurized process Swiss cheese	sodium citrate, sodium phosphate, sorbic acid, citric acid, lecithin
	Roast beef	sodium phosphate
	Roasted boneless turkey breast	sodium phosphate
	Roasted chicken breast with rib meat	spices, paprika, MSG, seasoning (hydrolyzed plant protein, disodium inosinate and guanylate), citric acid
	Shredded Cheddar cheese	microcrystalline cellulose, calcium chloride, ß-apo-8-carotenal (artificial colour), characteristic milk Cheddar artificial flavour
	Shredded mild Cheddar cheese	microcrystalline cellulose, artificial colour

FOOD TYPE	FOOD ITEM	ADDITIVES
Condiments	Arby's sauce	modified starch, karaya gum, citric acid, spices, sodium benzoate
	Au jus beef seasoning dip	caramel, hydrolyzed plant protein, MSG, spices, gelatin, disodium guanylate and inosinate, natural beef flavour, citric acid
	Bacon bits	sodium phosphate, sodium erythorbate, hickory smoke flavour, sodium nitrite
	Barbecue sauce	hickory smoke flavour, food starch, spices
	Cholesterol-free mayonnaise	xanthan gum, cellulose gel (microcrystalline cellulose), polysorbate 60, cellulose gum, natural flavour, paprika, calcium disodium EDTA
	Croutons	butter flavour
	Dill pickles	salt alum, sodium benzoate, natural spice flavours, FD&C Yellow #5 (tartrazine), polysorbate 80
	Horsey sauce	spices, locust bean gum, artificial flavour, calcium disodium EDTA
	Ketchup	spices

FOOD TYPE	FOOD ITEM	ADDITIVES
Condiments *cont'd*	Mayonnaise	spices and seasonings, calcium disodium EDTA
	Ranch dressing	spices, gums (arabic, xanthan, guar)
	Reduced-calorie honey mayonnaise	modified starch, MSG, spices, xanthan gum, caramel colour, natural flavour, FD&C Yellow #5 (tartrazine)
	Sour cream	modified starch, gelatin, mono- and diglycerides, citric acid, sodium caseinate, lactic acid, propylene glycol monoesters, guar gum, artificial flavour, carrageenan, monopotassium phosphate, potassium sorbate
	Sub dressing	spices, xanthan gum, propylene glycol alginate, artificial colour, hydrolyzed vegetable protein, calcium disodium EDTA
	Tartar sauce	spices and spice extractives, gums (arabic, guar, xanthan), calcium disodium EDTA
Side Orders	Curly fries	modified food starch, spices, cellulose gum, resin paprika, sodium acid pyrophosphate, natural flavour
	French fries	disodium dihydrogen pyrophosphate
	Potato cakes	natural flavour, disodium dihydrogen pyrophosphate

FOOD TYPE	FOOD ITEM	ADDITIVES
Salad dressings	Blue cheese	xanthan gum, natural flavour
	Buttermilk ranch	MSG, natural flavour, polysorbate 60, xanthan gum, sodium benzoate, spice, lactic acid, calcium disodium EDTA
	Honey French	paprika, spices, natural flavours, xanthan gum, propylene glycol, alginate
	Lite Italian	xanthan gum, spices, calcium disodium EDTA, FD&C Yellow #5 (tartrazine) and Yellow #6
	Thousand Island	spice, propylene glycol, alginate, paprika, natural flavour, oleoresin paprika, calcium disodium EDTA
Soups	Boston clam chowder	modified starch
	Chicken noodle	chicken seasoning (hydrolyzed plant protein, MSG, spice, caramel colour, disodium inosinate and guanylate), modified starch, spices
	Corn chowder	modified starch, MSG, spices, annatto
	Cream of broccoli	processed Cheddar cheese (sodium citrate, sodium phosphate, rennet and/or pepsin, calcium chloride, citric acid, sorbic acid, colour), modified starch, MSG, spices, chicken seasoning (hydrolyzed plant protein, MSG, spice)

FOOD TYPE	FOOD ITEM	ADDITIVES
Soups *cont'd*	French onion	seasoning, hydrolyzed plant protein, gelatin, caramel, lecithin, spices
	Lumberjack mixed vegetable	seasoning (hydrolyzed plant protein, MSG, modified starch), caramel, disodium guanylate and inosinate, natural and artificial flavours, paprika), modified starch, spices
	Split pea with ham	ham (cured with sodium phosphates, sodium erythorbate, sodium nitrite), MSG, spices, smoke flavour
	Tomato Florentine	seasoning (hydrolyzed plant protein)
	Beef with vegetables and barley	seasoning salt, hydrolyzed plant protein, modified starch, MSG, caramel, disodium guanylate and inosinate, emulsifier (propylene glycol monoesters, monoglycerides)
Breakfast items	Biscuit	sodium bicarbonate, sodium aluminum phosphate, natural flavour, monocalcium phosphate, citric acid
	Blueberry muffin	enriched bromated bleached flour (potassium bromate), citric acid, modified corn starch
	Cinnamon nut danish	sodium stearoyl lactylate, cinnamon, natural and artificial flavours, mono- and diglycerides, calcium proprionate, yellow colour (annatto and turmeric), potassium sorbate, agar

FOOD TYPE	FOOD ITEM	ADDITIVES
Breakfast items *cont'd*	Croissant	dough conditioner (diacetyl tartaric esters of mono- and diglycerides, ascorbic acid, potassium bromate, fungal amylase)
	Toastix	French bread (calcium sulfate, ammonium sulfate, potassium bromate), dough conditioner (contains mono- and diglycerides, ethoxylated mono- and diglycerides), calcium proprionate, modified starch, gum arabic, leavening (monocalcium phosphate, sodium bicarbonate), natural and artificial flavours, polysorbate
	"Maple" syrup	algin derivative, natural and artificial flavours, sodium benzoate, sorbic acid, sodium citrate, citric acid, caramel colour
	Sausage	spices, propyl gallate, citric acid, BHT
Desserts	Apple turnover	modified starch, citric acid, cinnamon
	Blueberry turnover	citric acid
	Cherry turnover	modified starch
	Chocolate chip cookie	chocolate chips (chocolate with lecithin, vanillin)

FOOD TYPE	FOOD ITEM	ADDITIVES
Beverages	Shake mix	artificial vanilla flavour, mono- and diglycerides, cellulose gum, guar gum, carrageenan, dipotassium phosphate, sodium citrate, sodium carbonate monohydrate
	Vanilla shake syrup	caramel colour, natural and artificial flavours, sodium benzoate, citric acid
	Chocolate shake syrup	caramel colour, sodium phosphate, vanillin, potassium sorbate
	Jamocha shake syrup	potassium sorbate, carrageenan, tetra-sodium pyrophosphate, vanillin
	Hot chocolate drink base	disodium phosphate, natural and artificial flavours, soy lecithin
Polar Swirls	Butterfinger (candy added to vanilla shake)	emulsifiers (glycerl-lacto esters of fatty acids and soy lecithin), glycerin, artificial flavour, FD&C Yellow #5 (tartrazine), TBHQ, citric acid
	Heath (candy added to jamocha shake)	chocolate (lecithin, vanillin), natural and artificial flavours, lecithin
	Oreo (cookie added to vanilla shake)	soy lecithin, vanillin

FOOD TYPE	FOOD ITEM	ADDITIVES
Polar Swirls *cont'd*	Reese's Peanut Butter Cup (candy added to chocolate shake)	milk chocolate (soy lecithin, vanillin), TBHQ, citric acid
	Snickers (candy added to vanilla shake)	milk chocolate (soy lecithin, vanillin), artificial flavour
Fats	Bun-toasting oil	lecithin, artificial flavour, artificial colour, TBHQ, citric acid, dimethylpolysiloxane
	Butter	colour
	Liquid margarine	TBHQ, citric acid, dimethylpolysiloxane
	Margarine/butter blend	vegetable mono- and diglycerides, vegetable lecithin, sodium benzoate, artificial flavour
	Vegetable frying oil	TBHQ, citric acid, dimethylpolysiloxane
	Vegetable frying shortening	monoglyceride citrate, propyl gallate, propylene glycol, dimethylpolysiloxane

* Arby's in Canada did not provide us with a list of ingredients. The following information is taken from a list of ingredients for their operations in the United States.

TABLE 19 Additives in Domino Pizza*

FOOD TYPE	FOOD ITEM	ADDITIVES
Crusts (ingredients vary from store to store)	Crust #1	mono- and diglycerides, sodium stearoyl-2-lactylate, artificial colour, cheese flavour (natural cheese flavour, sodium citrate, disodium guanylate and inosinate, spices)
	Crust #3	mono- and diglycerides, sodium caseinate, monocalcium phosphate, ascorbic acid, potassium bromate, natural and artificial flavours, lecithin, TBHQ
	Crust #4	mono- and diglycerides, sodium stearoyl lactylate, citric acid
Toppings	Beef for pizza	seasonings (sodium tripolyphosphate, spices, disodium guanylate and inosinate, BHA, BHT, citric acid)
	Black olives	ferrous gluconate
	Bold 'n' Spicy Italian sausage	seasonings (spices, sodium tripolyphosphate, caramel colour, disodium guanylate and inosinate, BHA, BHT, citric acid)
	Breakfast sausage	seasonings (spices, MSG, BHA, BHT, citric acid)
	Canadian-style bacon	sodium phosphate, sodium nitrite
	Canned mushrooms	may contain citric acid
	Cheese	rennet, calcium chloride, microcrystalline cellulose (less than 1%), sodium citrate, natural flavour, sodium propionate

FOOD TYPE	FOOD ITEM	ADDITIVES
Toppings *cont'd*	Chunked tomatoes	citric acid, calcium chloride
	Green chilies	citric acid, calcium chloride
	Green olives	lactic acid
	Green peppers	citric acid, erythorbic acid, calcium sulfate, and/or calcium phosphate
	Ham	sodium phosphate, hydrolyzed plant protein, sodium ascorbate, sodium nitrite
	Hot banana peppers	calcium chloride, FD&C Yellow #5 (tartrazine), sodium benzoate, polysorbate 80, turmeric, calcium disodium EDTA
	Italian sausage	seasonings (spices, sodium tripolyphosphate, caramel colour, disodium guanylate and inosinate, BHA, BHT, citric acid)
	Jalapeno peppers	calcium chloride, may contain citric acid
	Pepperoni	spices, lactic acid starter culture, oleoresin of paprika, hydrolyzed plant protein, sodium nitrite, BHA, BHT, citric acid
	Sauce	garlic powder (may contain citric acid and an anti-caking agent)

* Domino's Pizza in Canada did not provide us with a list of ingredients. The following information is taken from a list of ingredients for their operations in the United States.

TABLE 20 Additives in Dairy Queen Foods

FOOD TYPE	FOOD ITEM	ADDITIVES
Meat	Brazier label homestyle patty	–
	J.D. Sweid chili	starch, monosodium glutamate
	Brazier label wieners 6", 7" & 10"	sodium erythorbate, sodium nitrate, wood smoke, modified corn starch, paprika
	Pre-cooked bacon (Schneiders)	sodium phosphate, sodium erythorbate, sodium nitrate, wood smoke
	Brazier label sharp Cheddar cheese	annatto, calcium chloride, carotene (beta), citric acid, lecithin, milk coagulating enzyme, pepsin, rennet, sodium bicarbonate, sodium citrate, sodium phosphate (tribasic), sorbic acid
	Brazier grilled chicken breast	–
	Brazier breaded chicken fillet	Breading: citric acid, monosodium glutamate, sodium carbonate, sodium bicarbonate, sodium lauryl sulphate, triethyl citrate. Frying oil: BHA, BHT, dimethylpolysiloxane

FOOD TYPE	FOOD ITEM	ADDITIVES
Meat *cont'd*	Brazier label fish fillet	calcium silicate, monosodium glutamate (hydrolyzed plant protein, autolyzed yeast), sodium hexametaphosphate, sodium tripolyphosphate, modified corn starch The following are permitted food additives which may or may not be present in the flour of this product: acetone, amylase, ammonium persulphate, ammonium chloride, ascorbic acid, azodicarbonamide, benzoyl peroxide, bromelain, calcium carbonate, calcium sulphate, chlorine, chlorine dioxide, dicalcium phosphate, glucoamylase, lactase, lipoxidase, l-cysteine (hydrochloride), magnesium carbonate, potassium aluminum sulphate, potassium bromate, protease, sodium aluminum sulphate, tricalcium phosphate, monocalcium phosphate
Ice cream toppings	Chocolate syrup	potassium sorbate
	Butterscotch fudge	disodium phosphate, tartrazine, sunset yellow, caramel, vegetable oil shortening (BHA, BHT and citric acid), flavour (sodium benzoate, acetic acid)

FOOD TYPE	FOOD ITEM	ADDITIVES
Ice cream toppings *cont'd*	Hot chocolate fudge	disodium phosphate, potassium sorbate, lecithin, citric acid (in the hydrogenated palm kernel oil); propylene glycol (in flavour)
	Chocolate flavoured cone dip	lecithin
	Cold chocolate fudge	algin, sodium phosphate, BHA, BHT and citric acid (in vegetable oil shortening); mono- and diglycerides, lecithin, disodium phosphate, potassium sorbate, propyl paraben; ethyl alcohol and propylene glycol (in flavour)
	Blueberry	sodium benzoate, locust bean gum, sodium citrate, citric acid; ethyl alcohol and sodium benzoate (in flavour)
	Rum and butter	alcohol (in flavour); sunset yellow FCF, sodium benzoate and citric acid (in colour)
	Brandied peach	locust bean gum, tragacanth gum, potassium sorbate, malic acid; propylene glycol, caramel colour, alcohol, sodium benzoate, ethyl acetate, tartrazine, sunset yellow FCF (in flavour)
	Caramel fudge	disodium phosphate
	Liquid malt and pump	propylene glycol, sorbic acid, caramel

FOOD TYPE	FOOD ITEM	ADDITIVES
Ice cream toppings *cont'd*	Marshmallow	sodium lauryl sulphate, triethyl citrate, may contain citric acid, lactic acid, sodium carbonate (in dried egg albumen), cream of tartar; propylene glycol (in flavour)
	Blackberry	locust bean gum, sodium citrate, sodium benzoate, citric acid
	Toasted butter pecan	sodium benzoate, caramel colour; propylene glycol, caramel (in flavour)
	Red raspberry	locust bean gum, sodium citrate, sodium benzoate, amaranth liquid, citric acid; propylene glycol and ethyl alcohol (in flavour)
	Cherry	locust bean gum, sodium citrate, sodium benzoate, amaranth liquid, caramel colour, citric acid; propylene glycol, malic acid and ethyl alcohol (in flavour)
	Banana syrup	sodium benzoate, citric acid; tartrazine, sunset yellow FCF, sodium benzoate, alcohol and propylene glycol (in flavour)
	Blending syrup	potassium sorbate, propylene glycol, alginate, tragacanth gum, polydimethylsiloxane, silica, stearate emulsifiers, sorbic acid (in anti-foam), citric acid

FOOD TYPE	FOOD ITEM	ADDITIVES
Ice cream toppings *cont'd*	Coffee syrup	sodium benzoate, citric acid; sorbic acid, glycerine and caramel colour (in flavour)
	Creme de Menthe syrup	sodium benzoate, citric acid; tartrazine, FD&C Blue #1, sodium benzoate and glycerine (in flavour)
	Licorice syrup	sodium benzoate, tartrazine, amaranth, brilliant blue, citric acid
	Lime syrup	sodium benzoate, citric acid; tartrazine, FD&C Blue #1 and sodium benzoate (in flavour)
	Maple fudge syrup	sodium benzoate, citric acid; sodium benzoate, alcohol, hydrochloric acid, sodium bicarbonate, caramel colour and propylene glycol (in flavour)
	Vanilla syrup	sodium benzoate, caramel colour, citric acid; propylene glycol and caramel (in flavour)
Mr. Misty	Cherry	potassium sorbate, citric acid, tartrazine liquid, amaranth; alcohol and triacetin (in flavour)
	Grape	potassium sorbate, citric acid, amaranth, brilliant blue; propylene glycol (in flavour)

FOOD TYPE	FOOD ITEM	ADDITIVES
Mr. Misty *cont'd*	Lemon/lime	potassium sorbate, citric acid, tartrazine, brilliant blue; alcohol, propylene glycol, tartrazine, sodium benzoate, citric acid, BHA and tocopherols (in flavour)
	Orange	potassium sorbate, citric acid, sunset yellow FCF, tartrazine; BHA, tocopherols, ethyl alcohol, citric acid (in flavour); sodium benzoate, BHA and citric acid (in clouding agent)
	Raspberry	potassium sorbate, citric acid, brilliant blue, amaranth; ethyl alcohol, propylene glycol, lactic acid and acetic acid (in flavour)
	Strawberry	potassium sorbate, citric acid, amaranth, sunset yellow FCF; propylene glycol and ethyl alcohol (in flavour)
Sauces	Brazier sauce	sorbic acid, citric acid, caramel colour, acetic acid; BHT (in flavour); paprika

TABLE 21 Additives in Harvey's and Swiss Chalet Foods (Cara Operations Limited)*

FOOD TYPE	FOOD ITEM	ADDITIVES
Meat and other protein products	Hamburgers, chicken, hot dogs, sausages, fish	artificial colour, artificial flavour, calcium sulphate, calcium lactate, guar gum, lactic acid, lecithin, monocalcium phosphate, natural smoke flavour, natural flavour, sodium acid pyrophosphate, sodium bicarbonate, sodium erythorbate, sodium nitrite, xanthan gum
Milk and dairy products	Milk, ice cream, shakes, cream, flavouring syrups and toppings	artificial flavour, calcium sulphate, carob bean gum, carboxymethyl cellulose, calcium chloride, carrageenan, citric acid, cellulose gum, colour, disodium phosphate, guar gum, lactic acid, lecithin, lipase, locust bean gum, microbial enzyme, mono- and diglycerides, polysorbate 80, pepsin, propylene glycol, potassium sorbate, rennet, sorbic acid, sodium citrate, sodium bicarbonate, sodium benzoate, xanthan gum
Vegetable and fruit products	French fries, fruit pies, onion rings, soups, fruit toppings	ascorbic acid, calcium chloride, carrageenan, cellulose gum, disodium phosphate, glycerine, hydroxymethylcellulose, lecithin, locust bean gum, mono- and diglycerides, monocalcium phosphate, polysorbate 80, potassium sorbate, potassium chloride, propylene glycol, sodium acid pyrophosphate, sodium benzoate, sodium bisulphite, sodium bicarbonate, sodium caseinate, xanthan gum

FOOD TYPE	FOOD ITEM	ADDITIVES
Desserts	Cakes, pies	agar agar, ammonium bicarbonate, bacterial culture, carob bean gum, calcium carrageenan, citric acid, cream of tartar, colour, disodium phosphate, erythrobic acid, flavour, gelatin, gum arabic, hydroxypropyl methylcellulose, lecithin, locust bean gum, potassium sorbate, propylene glycol alginate, polysorbate 60, propylene glycol esters, rennet, shellac, sodium alginate, sodium benzoate, sorbitan, sodium bicarbonate, sodium citrate, tartaric acid, vanillin, xanthan gum
Fats and oils		artificial flavour, BHA, BHT, colour, dimethylpolysiloxane, lecithin, monoglyceride citrate
Condiments	Salad dressings, dipping sauces, jam, peanut butter, gravy and sauces, sandwich garnishes	alum, artificial flavour, BHA, BHT, citric acid, colour, cellulose gum, disodium guanylate, disodium inosinate, disodium EDTA, gum arabic, guar gum, mono- and diglycerides, microcrystalline cellulose, natural flavour, potassium sorbate, polysorbate 80, pectin, propylene glycol alginate, polysorbate 60, phosphoric acid, sodium acetate, sorbic acid, sodium benzoate, sodium citrate, xanthan gum

FOOD TYPE	FOOD ITEM	ADDITIVES
Bakery products	Hamburger buns, kaiser buns, hot dog buns, sandwich bread, muffins	azodicarbonamide, ammonium chloride, calcium carbonate, calcium sulphate, calcium iodate, calcium propionate, l-cysteine hydrochloride, monocalcium phosphate, polyoxyethylene (8) stearate, potassium bromate, potassium sorbate, polyoxyethylene (20) sorbitan monostearate, protease, sodium stearoyl-2-lactylate
Beverages	Soft drinks, Clamato juice	aspartame, caffeine, citric acid, monosodium glutamate, natural flavour, phosphoric acid, potassium sorbate, sodium benzoate, sodium citrate

* Cara Operations Limited provided us with information on food additives used in their products under general food categories rather than specific food items. In addition, they listed the additives alphabetically rather than by quantity used in each instance. No distinctions were made between products served at Harvey's and those served at Swiss Chalet operations.

TABLE 22 Additives in Restaurant Beverages

BEVERAGE	ADDITIVES
Pepsi-Cola	phosphoric acid, citric acid, caffeine, sodium benzoate, natural flavour
7-Up	citric acid, sodium citrate, sodium benzoate, natural flavours
Diet Pepsi	aspartame (contains phenylalanine), phosphoric acid, citric acid, sodium benzoate, caffeine, natural flavours, dimethylpolysiloxane
Orange Crush	citric acid, sodium benzoate, natural flavours, colour
Hires Root Beer	natural flavour, artificial flavour, caramel, phosphoric acid, sodium benzoate
Coca-Cola[1]	caramel colour, phosphoric acid, natural flavours, caffeine
Diet Coke[1]	caramel colour, aspartame, phosphoric acid, natural flavours, citric acid, sodium benzoate, caffeine, dimethylpolysiloxane
Sprite[1]	citric acid, natural lemon and lime flavours, sodium citrate, sodium benzoate

[1] Coca-Cola Company products

14

ALLERGY AWARE

IN OCTOBER 1991, the Canadian Restaurant and Foodservices Association launched a program called "Allergy Aware." According to Robin Garrett, Director of Research and Communication, the objective of the program is to help the public identify which restaurants provide ingredient information. In order to qualify for the program, operators must provide ingredient information in one or more of the following ways:

- a food allergy sensitivity chart, which specifically lists the additives MSG, tartrazine (U.S. yellow dye #5) and sulphites
- complete ingredient information on three or more main menu items
- complete ingredient information from three or more prepackaged meals

Restaurants participating in the program display an "Allergy Aware" decal on their windows and a poster welcoming ingredient inquiries on their premises. A trained staff member on each shift is available to respond to inquiries.

At present, forty companies and more than 1,800 locations are involved in the program. The following national chains participate in the Allergy Aware program in all of their locations across Canada:

- McDonald's Restaurants of Canada — 631 locations
- Mr. Submarine Ltd. — 362 locations
- Harvey's Restaurants — 250 locations
- Burger King Canada Inc. — 182 locations
- Swiss Chalet Restaurants — 129 locations
- Red Lobster Canada — 70 locations
- Cultures Fresh Food Restaurants — 65 locations
- Keg Restaurants — 61 locations
- Steak and Burger Restaurants — 29 locations

A growing number of restaurants participate in this program as well. For more information on the program and an up-to-date list of participants and restaurant locations, contact: Canadian Restaurant and Foodservices Association, Allergy Aware Program, 316 Bloor St. W., Toronto, ON, M5S 1W5; 416-923-8416 or toll free 1-800-387-5649.

15

FOR MORE INFORMATION

GROUPS

Environmental Health

Advocacy Group for the Environmentally Sensitive
1887 Chaine Ct.
Orleans, ON K1C 2W6
Tel: 613-830-5722

Allergy Asthma Information Association
65 Tromley Dr., #10
Islington, ON M9B 5Y7
Tel: 416-244-9312

Allergy Foundation of Canada
PO Box 1904
Saskatoon, SK S7K 3S5
Tel: 306-373-7591

Allergy and Environmental Health Association
5 Tangreen Ct., #704
North York, ON M2M 3Z1
Tel: 416-512-2628

Association des allergologues et immunologues du Québec
2, Complexe Desjardins, #3000
Montréal, PQ H5B 1G8
Tel: 514-845-2474

Consumers Health Organization of Canada
250 Sheppard Ave. E., #205
PO Box 248
Willowdale, ON M2N 5S9
Tel: 416-222-6517

Health Action Network Society
5262 Rumble St., #202
Burnaby, BC V5J 2B6
Tel: 604-435-0512

National Environmental Health Association
720 South Colorado Blvd., South Tower, #970
Denver, CO 80222-1925
U.S.A.
Tel: 303-756-9090

Parents of the Environmentally Sensitive
PO Box 434, Station R
Toronto, ON M4G 4C3
Tel: 416-424-1611

Consumer Protection

Consumers' Association of Canada
49 Auriga Dr.
Nepean, ON K2E 8A1
Tel: 613-723-0187

International Organization of Consumers Unions
Emmastraat 9,
2595 EG The Hague
The Netherlands
Tel: 31-70-347-63-31

Food Industry

Association des manufacturiers de produits alimentaires du Québec
200, rue MacDonald, #304
St-Jean-sur-Richelieu, PQ J3B 8J6
Tel: 514-349-1521

Bakery Council of Canada
885 Don Mills Rd., #301
Don Mills, ON M3C 1V9
Tel: 416-510-8041

Breakfast Cereal Manufacturers of Canada
885 Don Mills Rd., #301
Don Mills, ON M3C 1V9
Tel: 416-510-8036

Canadian Cattlemen's Association
6715 8th St. N.E., #215
Calgary, AB T2E 7H7
Tel: 403-275-8558

Canadian Chemical Producers' Association
350 Sparks St., #805
Ottawa, ON K1R 7S8
Tel: 613-237-6215
Toll Free: 1-800-267-6666

Canadian Council of Grocery Distributors
Place du Parc
PO Box 1082
Montreal, PQ H2W 2P4
Tel: 514-982-0267

Canadian Federation of Independent Meat Grocers
101 Duncan Mill Rd., #302
Don Mills, ON M3B 1Z3
Tel: 416-449-1976

Canadian Health Food Association
1093 West Broadway, #102A
Vancouver, BC V6H 1E2
Tel: 604-731-4664

Canadian Meat Council
5233 Dundas St. W.
Islington, ON M9B 1A6
Tel: 416-239-8411

Canadian Poultry and Egg Processors Council
1 Eva Rd., #300
Etobicoke, ON M9C 4Z5
Tel: 416-622-8621

Canadian Restaurant and Foodservices Association
316 Bloor St. W.,
Toronto, ON M5S 1W5
Tel: 416-923-8416

Canadian Soft Drink Association
55 York St., #330
Toronto, ON M5J 1R7
Tel: 416-362-2424

Canadian Sugar Institute
10 Bay St.
Toronto, ON M5C 1A2
Tel: 416-368-8091

Canadian Wine Institute
89 The Queensway W., #215
Mississauga, ON L5B 2V2
Tel: 905-273-5610

Coffee Association of Canada
885 Don Mills Rd., #301
Don Mills, ON M3C 1V9
Tel: 416-510-8024

Confectionary Manufacturers Association of Canada
885 Don Mills Rd., #301
Don Mills, ON M3C 1V9
Tel: 416-510-8034

Dairy Bureau of Canada
20 Holly St., #400
Toronto, ON M4S 3B1
Tel: 416-485-4453

Flavour Manufacturers
Association of Canada
885 Don Mills Rd., #301
Don Mills, ON M3C 1V9
Tel: 416-510-8036

Food Institute of Canada
1600 Scott St., #415
Ottawa, ON K1Y 4N7
Tel: 613-722-1000

Fresh for Flavour Foundation
1101 Prince of Wales Dr., #310
Ottawa, ON K2C 3W7
Tel: 613-226-4187

Grocery Products
Manufacturers of Canada
885 Don Mills Rd., #301
Don Mills, ON M3C 1V9
Tel: 416-510-8024

Institute of Edible Oil Foods
885 Don Mills Rd., #301
Don Mills, ON M3C 1V9
Tel: 416-510-8036

National Dairy Council
of Canada
141 Laurier Ave. W., #704
Ottawa, ON K1P 5J3
Tel: 613-238-4116

Tea Association of Canada
885 Don Mills Rd., #301
Don Mills, ON M3C 1V9
Tel: 416-510-8649

Irradiation of Food

Canadian Coalition to Stop
Food Irradiation
5262 Rumble St., #202
Burnaby, BC V5J 2B6
Tel: 604-435-0512

Canadian Institute for
Radiation Safety
555 Richmond St. W., #1106
Toronto, ON M5V 3B1
Tel: 416-366-6565

Consumers United for
Food Safety
PO Box 22928
Seattle, WA 98122
U.S.A.
Tel: 206-747-2659

Consumers United to Stop Food Irradiation
RR #1
Ilderton, ON N0M 2A0
Tel: 519-666-2072

Estrie contre l'irradiation
1865, Mitchell
Lennoxville, PQ J1M 2A3

Food & Water Incorporated
225 Lafayette St., #613
New York City, NY 10012
U.S.A.
Tel: 212-941-9340

Health Action Network Society
Food Irradiation Alert Committee
5262 Rumble St., #202
Burnaby, BC V5J 2B6
Tel: 604-435-0512

Nuclear Awareness Project
PO Box 2331
Oshawa, ON L1H 7V4
Tel: 416-725-1565

Probe International
225 Brunswick Ave.
Toronto, ON M5S 2M6
Tel: 416-964-9223

Organic Farming and Agriculture

Agricultural Institute of Canada
151 Slater St., #907
Ottawa, ON K1P 5H4
Tel: 613-232-9459

Back to the Farm Research Foundation
PO Box 69
Davidson, SK S0G 1A0

Canadian Farm Animal Care Trust
47 Jonathan St.
Uxbridge, ON L9P 1B7
Tel: 416-852-5581

Canadian Federation of Agriculture
75 Albert St., #1101
Ottawa, ON K1P 5E7
Tel: 613-236-3633

Canadian Organic Growers Inc.
PO Box 6408, Station J
Ottawa, ON K2A 3Y6
Tel: 613-395-5392

Canadian Organic Producers
Marketing Co-op
PO Box 2000
Girvin, SK S0G 1X0

Canadians for the Ethical
Treatment of Food Animals
PO Box 35597, Station E
Vancouver, BC V6M 4G9

Centre for Sustainable
Agriculture
Box 9, Group 15
Hadashville, MB R0E 0X0

City Farmer – Canada's Office
of Urban Agriculture
318 Homer St., #801
Vancouver, BC V6B 2V2
Tel: 604-685-5832

Crop Protection Institute of
Canada
21 Four Seasons Pl., #627
Etobicoke, ON M9B 6J8
Tel: 416-622-9771

Ecological Agriculture
Projects
Macdonald Campus, McGill
University
111 Lakeshore Rd., #21
PO Box 191
Ste-Anne-de-Bellevue, PQ
H9X 3V9
Tel: 514-398-7771

Heritage Seed Program
RR #3
Uxbridge, ON L9P 1R3
Tel: 416-852-7965

Institute for Alternative
Agriculture, Inc.
9200 Edmonston Rd., #117
Greenbelt, MD 20770
U.S.A.
Tel: 301-441-8777

International Federation of
Organic Agriculture
Movements
c/o Okozentrum Imsbach
D-6695 Tholey-Theley,
Germany
Tel: +49-6853-5190

Jubilee Centre for
Agricultural Research
115 Woolwich St.
Guelph, ON N1H 3V1
Tel: 519-837-1620

National Organic Farmers Association
RFD 2, Sheldon Rd.
Barre, MA 01005
U.S.A.
Tel: 508-355-2853

Organic Crop Improvement Association (International)
3185 Township Rd.
Bellefontaine, OH 43311
U.S.A.

Organic Crop Improvement Association – New Brunswick
RR #5
Debec, NB E0J 1J0

Organic Crop Improvement Association – Nova Scotia
PO Box 802
Kentville, NS B4N 4H8

Organic Crop Improvement Association (Ontario) Inc.
475 Laurier Ave., #1002
Ottawa, ON K1R 7X1
Tel: 613-234-8927

Organic Food Production Association of North America
c/o Alternatives Natural Food Market
453 Reynolds St.
Oakville, ON L6J 3M6
Tel: 905-844-2375

Organic Producers Marketing Cooperative
PO Box 2000
Girvin, SK S0G 1X0
Tel: 306-567-2810

Preservation of Agricultural Lands Society
261 King St.
Niagara-on-the-Lake, ON L1S 1J0
Tel: 905-468-2841

Sustainable Agriculture Association
10920 - 88 Avenue
PO Box 1063
Edmonton, AB T6G 0Z1
Tel: 403-955-2851

Women for the Survival of Agriculture
RR #1
Metcalfe, ON K0A 2P0

Packaging Industry

Canadian Packaging and Printing Machinery Manufacturers' Association
116 Albert St., #701
Ottawa, ON K1P 5G3
Tel: 613-232-7213

Canadian Plastics Institute
1262 Don Mills Rd., #48
Don Mills, ON M3B 2W7
Tel: 416-441-3222

Council on Plastics in the Environment
1001 Connecticut Ave. N.W., #401
Washington, DC 20036
U.S.A.
Tel: 202-331-0099

Packaging Association of Canada
2255 Sheppard Ave. E., #330
Willowdale, ON M2J 4Y1
Tel: 416-490-7860

Pesticides

Alberta Pesticide Action Network
PO Box 6117
Hinton, AB T7V 1X5
Tel: 403-865-7549

Canadian Association of Chemical Distributors
505 Consumers Road, #101
Willowdale, ON M2J 4V8
Tel: 416-502-1166

Canadian Association of Pesticide Control Officials
780 Blanshard St.
Victoria, BC V8W 2H1
Tel: 604-721-3118

Canadian Pest Management Society
Agriculture Canada
PO Box 1000
Aggassiz, BC V0M 1A0
Tel: 604-796-2221

International Group of
National Associations of
Manufacturers of
Agrochemical Products
79a, av Albert Lancaster
B-1180 Brussels, Belgium
Tel: 02-375-68-60

International Organization for
Biological Control of
Noxious Animals & Plants
c/o CSIRO Biological
Control Unit
335, av Paul-Parguel
F-34100 Montpellier, France
Tel: 33-67545140

National Coaltion Against
the Misuse of Pesticides
701 East St. S.E.
Washington, DC 20003
U.S.A.
Tel: 202-543-5450

Northwest Coalition for
Alternatives to Pesticides
PO Box 1393
Eugene, OR 97440,
U.S.A.
Tel: 504-344-5044

FOOD COOPERATIVES AND TRADING COMPANIES

Bridgehead Distributing/Tools
for Peace
9328 Jasper Ave.
Edmonton, AB T5H 3T5

Bridgehead Inc.
1011 Bloor St. W.
Toronto, ON M6H 1M1
Tel: 416-535-5070

Bridgehead Inc.
20 James St.
Ottawa, ON K2P 0T6
Tel: 613-567-1455;
1-800-565-8563

Earth Harvest Cooperative
10th Street Northwest, #102
Calgary, AB T2N 1V3

Harvest Collective
877 Westminster Ave.
Winnipeg, MB R3G 1B3

High Level Foods Co-op/
Bridgehead Retailer
10313 - 82nd Ave.
Edmonton, AB T6E 1Z9

Ontario Federation of Food
Cooperatives and Clubs
22 Mowat Ave.
Toronto, ON M6K 3E8
Tel: 416-533-7989;
416-588-3230

PSC Natural Foods
836 Viewfield Rd.
Victoria, BC V9A 4V1

Wild West Organic Harvest
Co-op
East 2471 Simpson Rd.
Richmond, BC V5X 2R2

FAST-FOOD RESTAURANT CONTACTS

McDonald's Food Information
McDonald's Restaurants of
Canada Limited
McDonald's Place
Toronto, ON M3C 3L4
Tel: 416-443-1000

Burger King Corporation
201 City Centre Dr., 8th Fl.
Mississauga, ON L5B 2T4
Tel: 905-273-5000

Wendy's Restaurants of
Canada Inc.
6715 Airport Rd., #301
Mississauga, ON L4V 1X2
Tel: 905-677-7023

Taco Bell
5600 Explorer Dr., #203
Mississauga, ON L4W 4Y2
Tel: 905-602-4924

A & W Food Services of
Canada Ltd.
171 West Esplanade, #300 N
Vancouver, BC V7M 3K9
Tel: 604-988-2141

Arby's (Canada) Inc.
6299 Airport Rd., #111
Mississauga, ON L4V 1N3
Tel: 905-672-2755

Domino's Pizza of Canada Inc.
500 Trillium Dr., #6
Kitchener, ON N2E 2K6
Tel: 519-748-1330

Harvey's
Cara Operations Ltd.
230 Bloor St. W.
Toronto, ON M5S 1T8
Tel: 416-962-4571

Swiss Chalet
Cara Operations Ltd.
230 Bloor St. W.
Toronto, ON M5S 1T8
Tel: 416-962-4571

Kentucky Fried Chicken
10 Carlson Ct., #300
Rexdale, ON M9W 6L2
Tel: 416-674-0367

Mr. Submarine
720 Spadina Ave., #300
Toronto, ON M5S 2T9
Tel: 416-962-6232

Pizza Hut
5600 Explorer Dr., 4th Fl.
Mississauga, ON L4W 4Y2
Tel: 905-602-8011

Dairy Queen
5245 Harvester Rd.
PO Box 430
Burlington, ON L7R 3Y3
Tel: 905-639-1492

FEDERAL GOVERNMENT

AGRICULTURE CANADA
Sir John Carling Building
930 Carling Ave.
Ottawa, ON K1A 0C5
Tel: 613-995-8963

Food Inspection Directorate
Agri-food Safety & Strategies Division
2255 Carling Ave.
Ottawa, ON K1A 0Y9
Tel: 613-995-5433

Food Inspection Directorate
Meat & Poultry Products Division
2255 Carling Ave.
Ottawa, ON K1A 0Y9
Tel: 613-995-5433

Food Inspection Directorate
Dairy, Fruit & Vegetable Division
2255 Carling Ave.
Ottawa, ON K1A 0Y9
Tel: 613-995-5433

Pesticides Directorate
Pesticides Information Call Line
Tel: 1-800-267-6315

CONSUMER AND CORPORATE AFFAIRS CANADA
Place du Portage
Hull (Ottawa, ON K1A 0C9)
Tel: 819-997-2938

Consumer & Corporate Affairs Regional Offices

1489 Hollis St., #1500
Halifax, NS B3J 3M5
Tel: 902-426-6080

Complexe Guy Favreau
200 René-Lévesque Ouest, Tour Est, #502
Montréal, PQ H2Z 1X4
Tel: 514-496-1797

4900 Yonge St., 6th Fl.
Willowdale, ON M2N 6B8
Tel: 416-224-4031

260 St. Mary Ave., #202
Winnipeg, MB R3C 0M6
Tel: 204-983-2366

800 Burrard St., #1400
Vancouver, BC V6Z 2H8
Tel: 604-666-5000

HEALTH AND WELFARE CANADA
Jeanne Mance Building
Ottawa, ON K1A 0K9
Tel: 613-957-2979

Health Protection Branch Regional Offices

1557 Hollis St.
Halifax, NS B3J 1V5
Tel: 902-426-2160

1001 Ouest boul. St-Laurent
Longueuil, PQ J4K 1C7
Tel: 514-238-5488

2301 Midland Ave.
Scarborough, ON M1P 4R7
Tel: 416-973-1451

510 Lagimodière Blvd.
Winnipeg, MB R2J 3Y1
Tel: 204-983-3004

220 - 4 Avenue S.E., #282
Calgary, AB T2P 3C3
Tel: 403-292-4650

3155 Willingdon Green
Burnaby, BC V5G 4P2
Tel: 604-666-3359

PROVINCIAL GOVERNMENTS

Alberta

Alberta Agriculture
7000 113 Street
Edmonton, AB T6H 5T6
Tel: 403-427-2127

Alberta Consumer &
Corporate Affairs
10025 Jasper Ave., 22nd Fl.
Edmonton, AB T5J 3Z5
Tel: 403-422-3935

Alberta Health
10025 Jasper Ave., 18th Fl.
Edmonton, AB T5J 2N3
Tel: 403-427-7164

British Columbia

Ministry of Agriculture,
Fisheries & Food
808 Douglas St., Parliament
Buildings
Victoria, BC V8W 2Z7
Tel: 604-356-2862

Ministry of Labour &
Consumer Services
1019 Wharf St.
Victoria, BC V8V 1X4
Tel: 604-387-3194

Ministry of Health
1515 Blanshard St., Station 5-2
Victoria, BC V8W 3C8
Tel: 604-387-2323

Manitoba

Manitoba Agriculture
809 Norquay Bldg.
401 York Ave.
Winnipeg, MB R3C 0P8
Tel: 204-945-3433

Manitoba Consumer &
Corporate Affairs
Consumers' Bureau
450 Broadway, #343
Winnipeg, MB R3C 0V8
Tel: 204-956-2040;
Toll free: 1-800-782-0067

Manitoba Health
599 Empress St.
PO Box 925
Winnipeg, MB R3C 2T6
Tel: 204-945-2818

New Brunswick

Department of Agriculture
Communications &
Education Branch
PO Box 6000
Fredericton, NB E3B 5H1
Tel: 506-453-2666

Department of Justice
Consumer Affairs Branch
Centennial Bldg., #476
Fredericton, NB E3B 5H1
Tel: 506-453-2719

Department of Health &
Community Services
PO Box 5100
Fredericton, NB E3B 5G8
Tel: 506-453-2536

Newfoundland

Department of Foresty &
Agriculture
Confederation Complex,
5th Fl.
PO Box 8700
St. John's, NF A1B 4J6
Tel: 709-729-3793

Department of Justice
Consumer Affairs Division
Confederation Bldg.
PO Box 8700
St. John's, NF A1B 4J6
Tel: 709-729-2566

Department of Health
Confederation Bldg., West
Block
PO Box 8700
St. John's, NF A1B 4J6
Tel: 709-729-0084

Northwest Territories

Department of Health
PO Box 1320
Yellowknife, NT X1A 2L9
Tel: 403-920-6173

Nova Scotia

Department of Agriculture
& Marketing
PO Box 190
Halifax, NS B3J 2M4
Tel: 902-424-3245

Department of Housing and Consumer Affairs
Consumer Affairs Division
PO Box 998
Halifax, NS B3J 2X3
Tel: 902-424-8946

Department of Health
Joseph Howe Bldg., 12th Fl., Hollis St.
PO Box 488
Halifax, NS B3J 2R8
Tel: 902-424-4391

Ontario

Ministry of Agriculture & Food
Consumer Information Centre
801 Bay St.
Toronto, ON M7A 2B2
Tel: 416-326-3400

Ministry of Consumer & Commercial Relations
Consumer Information Centre
555 Yonge St., Main Fl.
Toronto, ON M7A 2H6
Tel: 416-326-8555

Ministry of Health
Communications & Information Branch
Hepburn Block, 8th Fl., Queen's Park
Toronto, ON M7A 1S2
Tel: 416-327-4343; 1-800-268-1153

Prince Edward Island

Department of Agriculture
PO Box 2000
Charlottetown, PE C1A 7N8
Tel: 902-368-4880

Department of Justice and Attorney General
Consumer Services Division
PO Box 2000
Charlottetown, PE C1A 7N8
Tel: 902-368-4580

Department of Health & Social Services
PO Box 2000
Charlottetown, PE C1A 7N8
Tel: 902-368-4935

Quebec

Ministère de l'Agriculture, des Pêcheries et de l'Alimentation
Communications
200A Chemin Sainte-Foy, 7e étage
Québec, PQ G1R 4X6
Tel: 418-643-2517

Ministère de la justice
Office de la protection du consommateur
400, blvd. Jean-Lesage, #450
Québec, PQ G1K 8W4
Tel: 418-643-1484

Ministère de la Santé et des services sociaux
Communications
1075, ch. Sainte-Foy, 15e étage
Québec City, PQ G1S 2M1
Tel: 418-643-7167

Saskatchewan

Department of Agriculture & Food
Communications Branch
Walter Scott Bldg.
3085 Albert St.
Regina, SK S4S 0B1
Tel: 306-787-6395

Department of Justice
Consumer Protection Branch
1874 Scarth St.
Regina, SK S4P 3V7
Tel: 306-787-7881

Department of Health
Communications
3475 Albert St.
Regina, SK S4S 6X6
Tel: 306-787-2743

Yukon Territory

Renewable Resources
Agriculture
PO Box 2703
Whitehorse, YT Y1A 2C6
Tel: 403-667-5838

Justice
Consumer Services
PO Box 2703
Whitehorse, YT Y1A 2C6
Tel: 403-667-5257

Health & Social Services
PO Box 2703
Whitehorse, YT Y1A 2C6
Tel: 403-667-3518

PART FOUR:

THE FOOD ADDITIVE INDEX

INDEX

THIS INDEX LISTS and states the purpose of all the additives permitted in food sold in Canada. Just a reminder – according to the Canadian *Food and Drug Regulations* the following ingredients are not considered to be food additives:

- salt
- sugar
- starch
- vitamins
- mineral nutrients
- amino acids
- spices
- seasonings
- flavouring preparations
- agricultural chemicals
- food-packaging materials
- veterinary drugs

With the exception of some suspect flavouring agents and enhancers, the above substances have not been included in this list. The sweeteners saccharin and cyclamate are permitted only in "table-top" preparations and similarly have not been included. Additives printed in **bold type** are considered to be of questionable safety according to one or more food additive source books.

ADDITIVE	PURPOSE
acacia gum (gum arabic)	texture-modifying agent glazing and polishing agent
acetic acid	pH-adjusting agent preservative
acetic anhydride	starch-modifying agent
acetone	carrier or extraction solvent
acetone peroxide	bleaching, maturing and dough conditioning agent
acetylated monoglycerides	texture-modifying agent glazing and polishing agent release agent
acetylated tartaric acid esters of mono- and diglycerides	texture-modifying agent
adipic acid	pH-adjusting agent starch-modifying agent
agar	texture-modifying agent
alcohol	(*see* ethyl alcohol)
algin	texture-modifying agent

ADDITIVE	PURPOSE
alginic acid	texture-modifying agent
alkanet	colouring agent
allura red (U.S. red dye no. 40)	colouring agent
alum	(*see* potassium aluminum sulphate)
aluminum ammonium sulphate	(*see* ammonium aluminum sulphate)
aluminum calcium silicate	(*see* calcium aluminum silicate)
aluminum metal	colouring agent
aluminum potassium sulphate	(*see* potassium aluminum sulphate)
aluminum sodium sulphate	(*see* sodium aluminum sulphate)
aluminum sulphate	firming agent starch-modifying agent miscellaneous agent
amaranth (U.S. red dye no. 2)	colouring agent
ammonium alginate	texture-modifying agent
ammonium aluminum sulphate	pH-adjusting agent firming agent

ADDITIVE	PURPOSE
ammonium bicarbonate	pH-adjusting agent
ammonium carbonate	pH-adjusting agent
ammonium carrageenan	texture-modifying agent
ammonium chloride	yeast food
ammonium citrate, mono- and dibasic	pH-adjusting agent sequestering agent
ammonium furcelleran	texture-modifying agent
ammonium hydroxide	pH-adjusting agent
ammonium persulphate	bleaching, maturing and dough-conditioning agent miscellaneous agent
ammonium phosphate, mono- and dibasic	pH-adjusting agent yeast food
ammonium salt of phosphorylated glyceride	texture-modifying agent
ammonium sulphate	yeast food
amylase	food enzyme
amyloglucosidase (glucoamylase, maltase)	food enzyme

ADDITIVE	PURPOSE
annatto	colouring agent
anthocyanin	colouring agent
ß-apo-8'-carotenal	colouring agent
arabinogalactan	texture-modifying agent
artificial colour	colouring agent
artificial flavour	flavouring agent
ascorbic acid (vitamin C)	bleaching, maturing and dough-conditioning agent preservative
ascorbyl palmitate	preservative
ascorbyl stearate	preservative
aspartame (Nutrasweet)	sweetener
azodicarbonamide	bleaching, maturing and dough-conditioning agent
baker's yeast glycan	texture-modifying agent
beeswax	glazing and polishing agent release agent
beet red	colouring agent

ADDITIVE	PURPOSE
benzoic acid	preservative
benzoyl peroxide	bleaching, maturing and dough-conditioning agent
benzyl alcohol	carrier or extraction solvent
BHA	(*see* butylated hydroxyanisole)
BHT	(*see* butylated hydroxytoluene)
bicarbonate of soda	(*see* sodium bicarbonate)
bovine rennet	food enzyme
brilliant blue FCF (U.S. blue dye no. 1)	colouring agent
bromelain	food enzyme
brominated vegetable oil	miscellaneous agent
n-butane	pressure-dispensing agent
2-butanone	(*see* methyl ethyl ketone)
butylated hydroxyanisole (BHA)	preservative
butylated hydroxytoluene (BHT)	preservative
1,3-butylene glycol	carrier or extraction solvent

ADDITIVE	PURPOSE
caffeine	miscellaneous agent
caffeine citrate	miscellaneous agent
calcium acetate	pH-adjusting agent
calcium alginate	texture-modifying agent
calcium aluminum silicate	anti-caking agent
calcium ascorbate	preservative
calcium carbonate (chalk)	miscellaneous agent pH-adjusting agent yeast food texture-modifying agent
calcium carrageenan	texture-modifying agent
calcium chloride	firming agent yeast food pH-adjusting agent
calcium citrate	texture-modifying agent pH-adjusting agent firming agent

ADDITIVE	PURPOSE
calcium citrate *cont'd*	sequestering agent yeast food
calcium disodium EDTA (calcium disodium ethylenediamine tetratacetic acid)	sequestering agent
calcium disodium ethylenediamine tetraacetic acid (calcium disodium EDTA)	(*see* calcium disodium EDTA)
calcium fumarate	pH-adjusting agent
calcium furcelleran	texture-modifying agent
calcium gluconate	firming agent pH-adjusting agent texture-modifying agent
calcium glycerophosphate	texture-modifying agent
calcium hydroxide (slaked lime)	pH-adjusting agent
calcium hypophosphite	texture-modifying agent
calcium iodate	bleaching, maturing and dough-conditioning agent

ADDITIVE	PURPOSE
calcium lactate	firming agent pH-adjusting agent yeast food miscellaneous agent
calcium oxide	pH-adjusting agent
calcium peroxide	bleaching, maturing and dough-conditioning agent
calcium phosphate, dibasic (dicalcium orthophosphate)	texture-modifying agent firming agent miscellaneous agent pH-adjusting agent yeast food
calcium phosphate, monobasic	yeast food firming agent sequestering agent pH-adjusting agent
calcium phosphate, tribasic (tricalcium phosphate)	anti-caking agent pH-adjusting agent sequestering agent yeast food

ADDITIVE	PURPOSE
calcium phosphate, tribasic *cont'd.*	texture-modifying agent miscellaneous agent
calcium phytate	sequestering agent
calcium propionate	preservative
calcium silicate	anti-caking agent miscellaneous agent
calcium sorbate	preservative
calcium stearate	anti-caking agent release agent
calcium stearoyl-2-lactylate	bleaching, maturing and dough-conditioning agent whipping agent
calcium sulphate	texture-modifying agent firming agent pH-adjusting agent yeast food miscellaneous agent
calcium tartrate	texture-modifying agent

ADDITIVE	PURPOSE
candelilla wax	glazing and polishing agent
canthaxanthin	colouring agent
caramel	colouring agent
carbon black	colouring agent
carbon dioxide	pressure-dispensing agent carrier or extraction solvent
carboxymethyl cellulose	(*see* sodium carboxymethyl cellulose)
carnauba wax	glazing and polishing agent
carob bean gum (locust bean gum)	texture-modifying agent
carotene	colouring agent
carrageenan	texture-modifying agent
castor oil	carrier or extraction solvent release agent
catalase	food enzyme
cellulase	food enzyme

ADDITIVE	PURPOSE
cellulose gum	(*see* sodium carboxymethyl cellulose)
cellulose, microcrystalline (microcrystalline cellulose)	texture-modifying agent miscellaneous agent
charcoal	colouring agent
chlorine (gas)	bleaching, maturing and dough-conditioning agent
chlorine dioxide	bleaching, maturing and dough-conditioning agent
chloropentafluoroethane	pressure-dispensing agent
chlorophyll	colouring agent
citric acid	preservative sequestering agent pH-adjusting agent miscellaneous agent
citrus red no. 2	colouring agent
cochineal	colouring agent
copper gluconate	miscellaneous agent
cream of tartar	(*see* potassium acid tartrate)

ADDITIVE	PURPOSE
l-cysteine (hydrochloride)	bleaching, maturing and dough-conditioning agent
dicalcium orthophosphate	(*see* calcium phosphate, dibasic)
dichloromethane	(*see* methylene chloride)
diglycerides	(*see* mono- and diglycerides)
dimethylpolysiloxane	antifoaming agent release agent
dioctylsodium sulfo-succinate	miscellaneous agent
dipotassium phosphate	(*see* potassium phosphate, dibasic)
disodium EDTA (disodium ethylenediamine tetraacetic acid)	sequestering agent
disodium ethylenediamine tetraacetic acid (disodium EDTA)	(*see* disodium EDTA)
disodium inosinate	flavour enhancer
disodium guanylate	flavour enhancer
disodium phosphate	(*see* sodium phosphate, dibasic)
epichlorohydrin	starch-modifying agent

ADDITIVE	PURPOSE
epsom salt	(*see* magnesium sulphate)
erythorbic (isoascorbic) acid	preservative
erythrosine (U.S. red dye no. 3)	colouring agent
ethanol	(*see* ethyl alcohol)
ethoxyquin	miscellaneous agent
ethyl acetate	carrier or extraction solvent
ethyl alcohol (ethanol)	carrier or extraction solvent
ethyl alcohol denatured with methanol	carrier or extraction solvent
ethyl ß-apo-8-carotenate	colouring agent
ethylene oxide	miscellaneous agent
fast green FCF (U.S. green dye no. 3)	colouring agent
ferrous gluconate	miscellaneous agent
ficin	food enzyme
flavour	flavouring agent
fumaric acid	pH-adjusting agent

ADDITIVE	PURPOSE
furcelleran	texture-modifying agent
gelatin	texture-modifying agent
glucanase	food enzyme
glucoamylase (amyloglucosidase, maltase)	food enzyme
gluconic acid	pH-adjusting agent
glucono delta lactone	pH-adjusting agent miscellaneous agent
glucose isomerase	food enzyme
glucose oxidase	food enzyme
glycerine	(*see* glycerol)
glycerol (glycerin)	humectant carrier or extraction solvent glazing and polishing agent
glyceryl diacetate	carrier or extraction solvent
glyceryl monoacetate (monoacetin)	miscellaneous agent

ADDITIVE	PURPOSE
glyceryl triacetate (triacetin)	miscellaneous agent carrier or extraction solvent
glyceryl tributyrate (tributyrin)	carrier or extraction solvent
glycine	sequestering agent
guaiac gum	(*see* gum guaicum)
guar gum	texture-modifying agent
gum arabic	(*see* acacia gum)
gum benzoin	glazing and polishing agent
gum guaicum (guaiac gum)	preservative
gum tragacanth	(*see* tragacanth gum)
hemicellulase	food enzyme
hexane	carrier or extraction solvent
hydrochloric acid	pH-adjusting agent starch-modifying agent

ADDITIVE	PURPOSE
hydrogen peroxide	starch-modifying agent miscellaneous agent preservative
hydrolyzed vegetable protein (contains MSG)	flavour enhancer
hydroxylated lecithin	texture-modifying agent
hydroxypropyl cellulose	texture-modifying agent
hydroxylpropyl methylcellulose	texture-modifying agent
indigotine	colouring agent
invertase	food enzyme
irish moss gelose	(*see* carrageenan)
iron oxide	colouring agent
iso-ascorbic acid	(*see* erythorbic acid)
isobutane	pressure-dispensing agent
isopropanol	(*see* isopropyl alcohol)
isopropyl alcohol (isopropanol)	carrier or extraction solvent

ADDITIVE	PURPOSE
karaya gum	texture-modifying agent
lactase	food enzyme
lactic acid	pH-adjusting agent
lactylated mono- and diglycerides	texture-modifying agent
lactylic esters of fatty acids	texture-modifying agent miscellaneous agent
lanolin	miscellaneous agent
lecithin	texture-modifying agent release agent preservative
lecithin citrate	preservative
lipase	food enzyme
lipoxidase	food enzyme
l-leucine	miscellaneous agent
locust bean gum	(*see* carob bean gum)
MSG	(*see* monosodium glutamate)

ADDITIVE	PURPOSE
magnesia	(*see* magnesium oxide)
magnesium aluminum silicate	miscellaneous agent
magnesium carbonate	pH-adjusting agent anti-caking agent release agent miscellaneous agent
magnesium chloride	texture-modifying agent miscellaneous agent
magnesium citrate	pH-adjusting agent
magnesium fumarate	pH-adjusting agent
magnesium hydroxide	pH-adjusting agent
magnesium oxide	anti-caking agent pH-adjusting agent
magnesium silicate	anti-caking agent glazing and polishing agent miscellaneous agent release agent

ADDITIVE	PURPOSE
magnesium stearate	anti-caking agent release agent miscellaneous agent
magnesium sulphate	pH-adjusting agent starch-modifying agent miscellaneous agent
malic acid	pH-adjusting agent
maltase	(*see* glucoamylase)
maltol	flavour enhancer
manganese sulphate	yeast food
mannitol	texture-modifying agent release agent sweetener
methanol	(*see* methyl alcohol)
methyl alcohol (methanol)	carrier or extraction solvent
methylcellulose	texture-modifying agent

ADDITIVE	PURPOSE
methylene chloride (dichloromethane)	carrier or extraction solvent
methyl ethyl cellulose	texture-modifying agent miscellaneous agent
methyl ethyl ketone (2-butanone)	carrier or extraction solvent
methyl-p-hydroxybenzoate	preservative
methyl paraben	(*see* methyl-p-hydroxybenzoate)
microcrystalline cellulose	(*see* cellulose, microcrystalline)
milk coagulating enzyme	food enzyme
mineral oil	glazing and polishing agent release agent miscellaneous agent
modified starches	starch-modifying agent
monoacetin	(*see* glyceryl monoacetate)
monoammonium glutamate	flavour enhancer
monocalcium phosphate	(*see* calcium phosphate, monobasic)

ADDITIVE	PURPOSE
monoglycerides	texture-modifying agent antifoaming agent humectant release agent
mono- and diglycerides	texture-modifying agent antifoaming agent humectant release agent carrier or extraction solvent
monoglyceride citrate	preservative carrier or extraction solvent
monoisopropyl citrate	preservative
monopotassium glutamate	flavour enhancer
monosodium glutamate (MSG)	flavour enhancer
monosodium phosphate	(*see* sodium phosphate, monobasic)
natamycin	preservative
natural colours	colouring agent

ADDITIVE	PURPOSE
natural flavours	flavouring agent
nitrate	(*see* potassium nitrate, sodium nitrate)
nitric acid	starch-modifying agent
nitrite	(*see* potassium nitrite, sodium nitrite)
nitrogen	pressure-dispensing agent
2-nitropropane	carrier or extraction solvent
nitrous oxide	pressure-dispensing agent
Nutrasweet	(*see* aspartame)
oat gum	texture-modifying agent
octenyl succinic anhydride	starch-modifying agent
octafluorocyclobutane	pressure-dispensing agent
orchil	colouring agent
oxystearin	miscellaneous agent
ozone	miscellaneous agent
pancreas extract	miscellaneous agent

ADDITIVE	PURPOSE
pancreatin	food enzyme
papain	food enzyme
paprika	colouring agent
paraffin wax	miscellaneous agent
pectin	texture-modifying agent
pectinase	food enzyme
pentosanase	food enzyme
pepsin	food enzyme
peracetic acid	starch-modifying agent
petrolatum	glazing and polishing agent miscellaneous agent
phosphoric acid	pH-adjusting agent sequestering agent yeast food
phosphorous oxychloride	starch-modifying agent
polydextrose	texture-modifying agent

ADDITIVE	PURPOSE
polyethylene glycol	antifoaming agent miscellaneous agent
polyglycerol esters of fatty acids	texture-modifying agent
polyglycerol esters of interesterified castor oil fatty acids	texture-modifying agent
polyoxyethylene (20) sorbitan monooleate	(*see* polysorbate 80)
polyoxyethylene (20) sorbitan monostearate	(*see* polysorbate 60)
polyoxyethylene (20) sorbitan tristearate	(*see* polysorbate 65)
polyoxyethylene (8) stearate	texture-modifying agent
polysorbate 60 (polyoxyethylene (20) sorbitan monostearate)	texture-modifying agent
polysorbate 65 (polyoxyethylene (20) sorbitan tristearate)	texture-modifying agent
polysorbate 80 (polyoxyethylene (20) sorbitan monooleate)	texture-modifying agent
polyvinylpyrrolidone	miscellaneous agent

ADDITIVE	PURPOSE
ponceau sx (U.S. red dye no. 4)	colouring agent
potassium acid tartrate (potassium bitartrate, cream of tartar)	pH-adjusting agent
potassium alginate	texture-modifying agent
potassium aluminum sulphate (alum)	firming agent pH-adjusting agent miscellaneous agent
potassium bicarbonate	pH-adjusting agent
potassium bisulphite	preservative
potassium bitartrate	(*see* potassium acid tartrate)
potassium bromate	bleaching, maturing and dough-conditioning agent
potassium carbonate	pH-adjusting agent
potassium carrageenan	texture-modifying agent
potassium chloride	pH-adjusting agent yeast food texture-modifying agent

ADDITIVE	PURPOSE
potassium citrate	pH-adjusting agent texture-modifying agent
potassium fumarate	pH-adjusting agent
potassium furcelleran	texture-modifying agent
potassium hydroxide (caustic potash)	pH-adjusting agent
potassium iodate	bleaching, maturing and dough-conditioning agent
potassium lactate	pH-adjusting agent
potassium metabisulphite	preservative
potassium nitrate (saltpetre)	preservative
potassium nitrite	preservative
potassium permanganate	starch-modifying agent
potassium persulphate	bleaching, maturing and dough-conditioning agent
potassium phosphate	(*see* potassium phosphate, monobasic)
potassium pyrophosphate	(*see* potassium pyrophosphate, tetrabasic)

ADDITIVE	PURPOSE
potassium phosphate, monobasic (potassium phosphate)	starch-modifying agent yeast food
potassium phosphate, dibasic (dipotassium phosphate)	texture-modifying agent yeast food pH-adjusting agent
potassium pyrophosphate, tetrabasic (potassium pyrophosphate)	sequestering agent
potassium sodium tartrate	(*see* sodium potassium tartrate)
potassium sorbate	preservative
potassium stearate	miscellaneous agent
potassium sulphate	pH-adjusting agent
potassium tartrate	pH-adjusting agent
propane	pressure-dispensing agent
1,2-propanediol	(*see* propylene glycol)
propionic acid	preservative
propyl gallate	preservative

ADDITIVE	PURPOSE
propyl paraben	(*see* propyl-p-hydroxy benzoate)
1,2-propylene glycol (1,2-proanediol)	anti-caking agent humectant carrier or extraction solvent
propylene glycol alginate	texture-modifying agent
propylene glycol ether of methylcellulose	(*see* hydroxypropyl methylcellulose)
propylene glycol mono fatty acid esters	texture-modifying agent
propylene glycol monoesters and diesters of fat-forming fatty acids	carrier or extraction solvent
propylene oxide	starch-modifying agent
propyl paraben	(*see* propyl-p-hydroxy benzoate)
propyl-p-hydroxy benzoate	preservative
protease	food enzyme
pullulanase	food enzyme
quillaia extract	miscellaneous agent

ADDITIVE	PURPOSE
rennet	food enzyme
riboflavin (vitamin B)	colouring agent
rochelle salts	(*see* sodium potassium tartrate)
saffron	colouring agent
saponin	miscellaneous agent
saunderswood	colouring agent
shellac	glazing and polishing agent
silicon dioxide	anti-caking agent
silver metal	colouring agent
smoke	flavouring agent
sodium acetate	pH-adjusting agent starch-modifying agent
sodium acid pyrophosphate	texture-modifying agent pH-adjusting agent sequestering agent
sodium acid tartrate	pH-adjusting agent

ADDITIVE	PURPOSE
sodium alginate	texture-modifying agent
sodium aluminum phosphate	pH-adjusting agent texture-modifying agent
sodium aluminum silicate	anti-caking agent
sodium aluminum sulphate	firming agent pH-adjusting agent miscellaneous agent
sodium ascorbate	preservative
sodium benzoate	preservative
sodium bicarbonate (bicarbonate of soda)	pH-adjusting agent starch-modifying agent miscellaneous agent
sodium bisulphate	pH-adjusting agent
sodium bisulphite	preservative
sodium carbonate	pH-adjusting agent starch-modifying agent miscellaneous agent

ADDITIVE	PURPOSE
sodium carboxymethyl cellulose	texture-modifying agent miscellaneous agent
sodium carrageenan	pH-adjusting agent
sodium cellulose glycolate	texture-modifying agent
sodium chlorite	starch-modifying agent
sodium citrate	texture-modifying agent sequestering agent miscellaneous agent pH-adjusting agent
sodium diacetate	preservative
sodium dithionite	preservative
sodium erythorbate	preservative
sodium ferrocyanide decahydrate	anti-caking agent miscellaneous agent
sodium fumarate	pH-adjusting agent
sodium furcelleran	texture-modifying agent

ADDITIVE	PURPOSE
sodium gluconate	texture-modifying agent pH-adjusting agent
sodium hexametaphosphate	texture-modifying agent whipping agent pH-adjusting agent sequestering agent miscellaneous agent
sodium hydroxide	pH-adjusting agent starch-modifying agent
sodium hypochlorite	starch-modifying agent
sodium iso-ascorbate	preservative
sodium lactate	pH-adjusting agent
sodium lauryl sulphate	whipping agent
sodium metabisulphite	preservative
sodium metaphosphate	(*see* sodium hexametaphosphate)
sodium methyl sulphate	miscellaneous agent
sodium nitrate (soda niter)	preservative

ADDITIVE	PURPOSE
sodium nitrite	preservative
sodium phosphate, monobasic (sodium phosphate)	texture-modifying agent pH-adjusting agent sequestering agent
sodium phosphate, dibasic (disodium phosphate)	texture-modifying agent pH-adjusting agent sequestering agent miscellaneous agent
sodium phosphate, tribasic (trisodium phosphate)	pH-adjusting agent texture-modifying agent
sodium phosphate	(*see* sodium phosphate, monobasic)
sodium polyphosphate	(*see* sodium tripolyphosphate)
sodium potassium copper chlorophyllin	miscellaneous agent
sodium potassium tartrate (rochelle salts)	texture-modifying agent pH-adjusting agent
sodium propionate	preservative

ADDITIVE	PURPOSE
sodium pyrophosphate, tetra-basic	texture-modifying agent pH-adjusting agent sequestering agent
sodium salt of methyl-p-hydroxy benzoic acid	preservative
sodium salt of propyl-p-hydroxy benzoic acid	preservative
sodium silicate	miscellaneous agent
sodium sorbate	preservative
sodium stearate	miscellaneous agent
sodium stearoyl-2-lactylate	bleaching, maturing and dough-conditioning agent texture-modifying agent whipping agent miscellaneous agent
sodium stearyl fumarate	bleaching, maturing and dough-conditioning agent
sodium sulphate	miscellaneous agent yeast food

ADDITIVE	PURPOSE
sodium sulphite	preservative miscellaneous bleaching, maturing and dough-conditioning agent
sodium tartrate	texture-modifying agent
sodium tetraphosphate	(*see* sodium pyrophosphate)
sodium thiophosphate	miscellaneous agent
sodium thiosulphate	miscellaneous agent
sodium trimetaphosphate	starch-modifying agent
sodium tripolyphosphate	texture-modifying agent sequestering agent pH-adjusting agent starch-modifying agent miscellaneous agent
sorbic acid	preservative
sorbitan monostearate	texture-modifying agent
sorbitan trioleate	texture-modifying agent
sorbitan tristearate	texture-modifying agent

ADDITIVE	PURPOSE
sorbitol	texture-modifying agent release agent humectant sweetener
spermaceti wax	glazing and polishing agent
stannous chloride	miscellaneous agent
stearic acid	release agent miscellaneous agent
stearyl citrate	sequestering agent
stearyl monoglyceridyl citrate	texture-modifying agent
succinic anhydride	starch-modifying agent
sucrose acetate isobutyrate (SAIB)	miscellaneous agent
sulphur dioxide	(*see* sulphurous acid)
sulphuric acid	pH-adjusting agent starch-modifying agent
sulphurous acid	pH-adjusting agent preservative

ADDITIVE	PURPOSE
sunset yellow FCF (U.S. yellow dye no. 6)	colouring agent
talc	miscellaneous agent
tannic acid	miscellaneous agent texture-modifying agent
tartaric acid	preservative pH-adjusting agent
tartrazine (U.S. yellow dye no. 5)	colouring agent
tetrasodium pyrophosphate	(*see* sodium pyrophosphate)
thaumatin	sweetener
titanium dioxide	colouring agent
tocopherols	preservative
tragacanth gum (gum tragacanth)	texture-modifying agent
triacetin	(*see* glyceryl triacetate)
tributyrin	(*see* glyceryl tributyrate)
tricalcium (ortho) phosphate	(*see* calcium phosphate, tribasic)

ADDITIVE	PURPOSE
triethyl citrate	carrier or extraction solvent whipping agent
trisodium phosphate	(*see* sodium phosphate, tribasic)
turmeric	colouring agent
urea	yeast food
wood smoke	preservative
xanthan gum	texture-modifying agent
xanthophyll	colouring agent
xylitol	sweetener
zein	glazing and polishing agent
zinc sulphate	yeast food

ACKNOWLEDGEMENTS

POLLUTION PROBE was established in 1969 as an independent, non-profit, research-based charitable organization. Through memberships and donations, more than 60,000 Canadians are Pollution Probe Partners. Part of its mandate is to produce educational materials on environmental issues. At Pollution Probe, PATTY CHILTON was responsible for the in-house co-ordination and production of *Additive Alert* and many other Pollution Probe publications.

RANDEE L. HOLMES is a freelance writer and researcher who has contributed to Pollution Probe's *The Canadian Green Consumer Guide* (McClelland & Stewart, 1989 and 1991), *The Canadian Junior Green Guide* (McClelland & Stewart, 1990), *The Canadian Green Calendar 1991* (McClelland & Stewart, 1990), *The Kitchen Handbook* (McClelland & Stewart, 1992), and *Profit from Pollution Prevention*. She is a candidate for the Master of Environmental Studies degree at York University, Toronto.

After having worked together successfully in the past on other Pollution Probe publications, Patty Chilton and Randee Holmes teamed up again to produce *Additive Alert*. Like all projects, however, *Additive Alert* is the result of *many* people's efforts. We would particularly like to thank Sarah Boon and Amish Morrell for their countless hours of hard work and dedication to the project. Randee would also like to thank Jerri Burke for providing a haven in which to write the book. Other volunteers and Pollution Probe staff who deserve special thanks are: Martin Baker, Gail Camilleri, Sheri Camilleri, Andy Carroll, Christopher Gladun, Janine Ferretti, Karen McMaster, Iori Miller, Steven Robinson, Ellen Schwartzel, Janet Sumner and Sharon Taylor. We also extend our gratitude to Dinah Forbes and Doug Gibson at McClelland & Stewart.

The following people participated in the review process for this book. They verified the technical accuracy of the material and provided valuable advice on the book's content. With regards to some issues, however, the opinions of the reviewers differed from those of Pollution Probe. Inclusion in this list in no way suggests that the book necessarily reflects the position of the reviewers or implies an endorsement on their part. We thank William Glenn; Dr. Ross Hume Hall; Dr. Stuart Hill (Ecological Agriculture Projects, McGill University); Linda Holmes; Irene Kock (Nuclear Awareness Project); Marjorie Lamb; Linda Pim; Dr. A. Venket Rao (Department of Nutritional Sciences, University of Toronto); Suzanne St. Jacques-Hamlin (Food Division, Consumer and Corporate Affairs); Toby Vigod; and Dr. K. Weiss (Evaluation Division, Bureau of Microbial Hazards, Health and Welfare Canada).

About Pollution Probe

Pollution Probe, founded in 1969, is an independent registered charitable organization and one of Canada's leading public environmental interest groups. It has been at the forefront of Canada's central environmental issues for over twenty-four years and is responsible for significant improvements in reducing air and water pollution, curbing the generation of solid and hazardous waste, implementing stricter control on the release of toxic chemicals into our environment, and empowering citizens with legal rights to a healthy environment.

Over the years, Pollution Probe has vigorously exercised its mandate to educate Canadian children, students, and adults about environmental problems and their solutions. This book is but one of a long list of publications that puts such knowledge directly into the hands of individuals. Others include *The Canadian Green Consumer Guide*; *The Junior Green Guide*; and *The Kitchen Handbook, an Environmental Guide*.

Pollution Probe is dedicated to achieving positive, tangible, and practical environmental change for the benefit of all Canadians through research, education, and advocacy.

Donations may be sent to:

Pollution Probe
12 Madison Avenue
Toronto, Ontario
M5R 2S1